LIFE CYCLES

Written by
Marilyn Marks

Cover Photography
© Dwight Kuhn

Editor
Joel Kupperstein

Illustrator
Kate Flanagan

Project Director
Carolea Williams

TABLE OF CONTENTS

Introduction 3

Monarch Butterfly
I Am a Butterfly 5
The Life of a Butterfly 7
Caterpillar Keeper 10

Bean
My Suitcase Is Packed 11
Bean Sprouts 12
Comparing Seeds 13

Sunflower
Sunflower Game 15
Model Sunflowers 18
Grow Your Own Sunflowers 20

Wood Frog
Wood Frog Life Cycle Wheel 22
Surviving a Cold Winter 25
If I Were a Frog Egg 27

Ladybug
Ladybugs Up Close 28
What an Appetite! 29
Ladybug, Ladybug, Fly away Home 32

Chicken
Investigating Eggs 35
I've Grown My Comb and Wattles 37
Simon Crows 39

Jumping Spider
What Is a Spider? 41
The Life of a Spider 43
Spider Web Math 45

Maple Tree
Beautiful Fall Leaves 47
Spinning Seeds 49
The Maple Tree's Sweet Sap 52

Green Snake
Simulating Snake Eggs 54
Snakeskin and Scales 55
Skinny as a Snake 57

Hummingbird
Tiny Nests Are Best 61
Fledglings Learn to Fly 63
Plant a Hummingbird Garden 65

Horse
A Foal Stands Up 66
Horse Tag 69
A Horse of a Different Color 70

Fighting Fish
Bubble Nests for Babies 73
I'm Just a Small Fry 74
Design a Better Betta 77

Putting It Together 79

INTRODUCTION

Life Cycles is the perfect companion to the *Life Cycles* series of nonfiction books for young readers. This resource guide provides a wealth of activities that focus on the plants and animals featured in the *Life Cycles* series.

The *Life Cycles* books invite children to explore the life stages of a variety of plants and animals, such as wood frogs, ladybugs, maple trees, and horses. Young readers explore up close such interesting aspects of nature as a butterfly's chrysalis, a majestic sunflower, the courtship of a jumping spider, and a molting green snake. The final pages of each book depict all the stages of the plant or animal's life cycle and challenge readers to put them in order.

Collect the following *Life Cycles* books to use with this activity guide.

Monarch Butterfly
Bean
Sunflower
Wood Frog
Ladybug
Chicken
Jumping Spider
Maple Tree
Green Snake
Hummingbird
Horse
Fighting Fish

These engaging science activities apply to specific *Life Cycles* books. They focus on either the entire life cycle of an animal or a plant or on specific details from the book. Three activities are presented for each of the twelve books in the series. Background information, a materials list, and easy-to-follow instructions are provided for each activity. As your students read the *Life Cycles* books and study living things, have them keep a journal that details what they learn. This journal can be anything from a spiral-bound notebook to a stack of writing paper stapled inside a decorated construction-paper cover. Each activity in this book concludes with questions or ideas for students to respond to in their journal. These response journals are valuable tools for a number of reasons—they help students internalize their learning by putting it in their own words; they offer students the opportunity to apply their learning or hypothesize about it; they provide a reference resource for students; they provide a springboard for whole-class or small-group discussion; and they provide an informal, authentic means of assessing students' learning.

Putting It Together
This resource concludes with suggestions for helping students synthesize what they learned from each *Life Cycles* book. Conduct these activities after completing the entire series, and guide students toward understanding the key life-science concept that all living things share the following characteristics:

- They grow and change.
- They take in nutrients.
- They give off waste.
- They reproduce.
- They are made of smaller structures that can be studied.
- They respond to stimuli.

Your students' interest in and knowledge of life science will grow as they participate in the meaningful activities provided in this resource. *Life Cycles* gives you everything you need to make science come alive for your students.

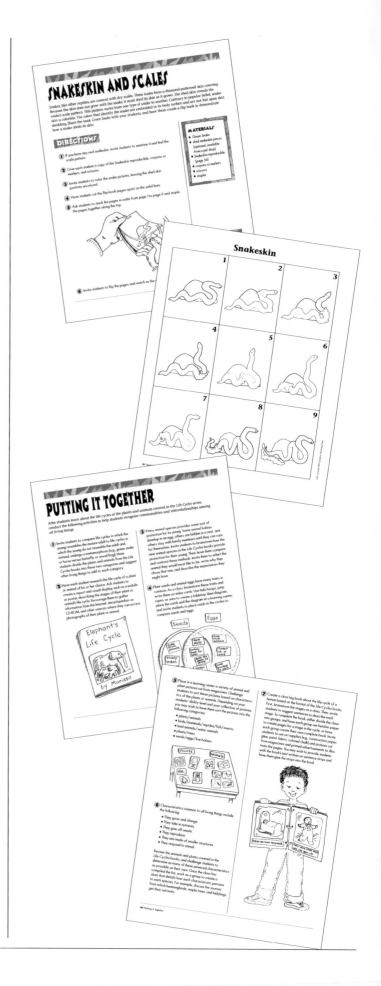

I AM A BUTTERFLY

How can you tell a butterfly from a moth? Butterflies have a thin body and slender antennae with knobs on the ends, and they rest with their wings folded up. Butterflies are active during the day, and they build a hanging chrysalis in which they undergo metamorphosis. Moths tend to have plump bodies and feathery antennae, and they rest with their wings open. Moths are mainly active at night and usually form their cocoon on the ground. Share the book *Monarch Butterfly* with your students so they can get a better look at this beautiful insect. Then, conduct the following activity in which students create a replica of a monarch butterfly.

DIRECTIONS

1 Give each student a copy of the Butterfly reproducible.

2 Have students cut out the butterfly patterns and the large, white spaces in the wings.

3 Ask them to hold the butterfly patterns together, printed sides out, and hole-punch the white circles near the edges of the wings.

4 Have them glue tissue paper between the patterns and trim the edges.

 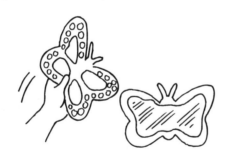

5 Invite students to tape a construction-paper strip to the underside of their butterfly and fold the wings slightly upward (away from the strip). Have them tape the paper strip around their wrist and move their arm up and down to simulate the flying motion of a butterfly. They can fold the wings up to show their butterfly resting.

MATERIALS

◆ *Monarch Butterfly*
◆ Butterfly reproducible (page 6)
◆ scissors
◆ hole punch
◆ glue
◆ orange tissue paper
◆ construction-paper strips
◆ transparent tape

IN YOUR JOURNAL

1. What does a butterfly look like? Draw a diagram.
2. Why do you think butterflies have colored wings?
3. Why do you think butterflies rest with their wings folded up?

Butterfly

THE LIFE OF A BUTTERFLY

A butterfly's life cycle includes many changes over a short period of time, called metamorphosis. This metamorphosis involves four steps. The monarch butterfly begins life as a very tiny egg. From this egg hatches the larva, or caterpillar. The caterpillar grows rapidly and sheds its skin several times. Then, the caterpillar changes into a pupa inside a "sleeping bag" called a chrysalis. When the chrysalis opens, an adult butterfly emerges. Then, the life cycle repeats itself. For most adult butterflies, life is very short, usually only a few weeks. However, the monarch butterfly lives for about two years. Share the book *Monarch Butterfly* with your students, and have them complete the activity below to learn about the butterflies' metamorphosis.

DIRECTIONS

1 Give each student a copy of the Butterfly Life Cycle reproducible and a piece of construction paper.

2 Have students color the leaf, branch, and adult butterfly and glue the page to the construction paper to create a colorful frame.

3 Give students sesame seeds, and ask them to glue several seeds onto the leaf to represent butterfly eggs.

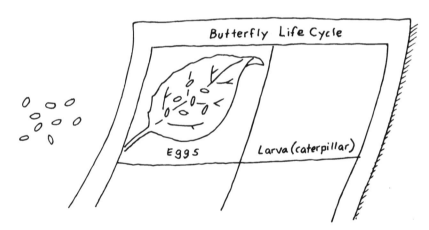

4 Hand out green chalk, a small piece of white paper, and a cotton ball to each student. Have students make a powdery chalk smudge on the paper, stretch out the cotton ball, and coat the cotton ball with the green chalk powder. Lightly spray each student's "chrysalis" with hair spray. Set aside to dry.

MATERIALS

- *Monarch Butterfly*
- Butterfly Life Cycle reproducible (page 9)
- large construction paper
- crayons or markers
- glue
- sesame seeds
- green chalk
- white paper
- cotton balls
- hair spray
- 6" (15 cm) black, white, and yellow pipe-cleaner pieces
- pencils

5 Demonstrate how to make a monarch caterpillar (its larval stage) by holding black, white, and yellow pipe-cleaner pieces against a pencil with your thumb and fingers, completely winding the pieces around the pencil with your other hand, and removing the pencil.

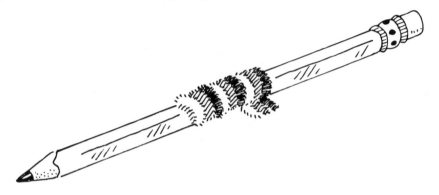

6 Give students pipe-cleaner pieces, and invite them to make their own larva to glue to the Larva section of the reproducible.

7 Have students glue the cotton-ball chrysalis to the Pupa section of the reproducible.

IN YOUR JOURNAL

1. Describe the four stages of the butterfly life cycle.
2. What happens inside a butterfly's chrysalis?
3. Compare a butterfly to a caterpillar.

Butterfly Life Cycle

Eggs

Larva (caterpillar)

Pupa (chrysalis)

Adult Butterfly

CATERPILLAR KEEPER

Caterpillars are very hungry creatures. The monarch butterfly caterpillar is very picky, too—it only eats milkweed plants. Caterpillars grow very quickly as a consequence of their generous diet. Because their outer covering, called an exoskeleton, does not grow or stretch, they literally "pop" out of their skin. The skin splits open, and the caterpillar crawls out, leaving the old skin behind. Share the pages of the book *Monarch Butterfly* with your students. Then, take them on a caterpillar walk, and see if you can bring some caterpillars back to your classroom in the colorful caterpillar keepers described below. Be sure to bring along leaves of the plants where you found them.

DIRECTIONS

1 Have each student cut a window in one side of a milk carton.

2 Invite students to use a variety of art supplies to decorate their carton. Be sure they are able to open and close the top of their carton.

3 Have students cut a piece of nylon netting to fit over the window and tape it in place.

4 Go on a caterpillar walk around the school, and have students bring their keepers to hold any caterpillars they might find. Caterpillars are more abundant in the warmer weather of spring and summer. They are most active from the late morning through the early afternoon. If the conditions at your school are unsuitable for collecting caterpillars, obtain them through a science supply house.

5 Let students observe the caterpillars in the wild first, and then assist them in picking off the leaf with the caterpillar. Have students open the top of their keeper and place the caterpillar and a few extra leaves inside.

6 Encourage students to observe the caterpillars over the next week or two. Have them record in their journal the number of leaves eaten, draw pictures of the caterpillars, and look for evidence of molting.

MATERIALS
- *Monarch Butterfly*
- milk cartons
- scissors
- art supplies (e.g., paint, glitter, tissue paper, glue)
- nylon netting
- transparent tape
- caterpillars (optional)

IN YOUR JOURNAL
1. Where did you find caterpillars?
2. Can you put any kind of leaf in with the caterpillars? Why or why not?
3. Why do caterpillars eat so much?

MY SUITCASE IS PACKED

What we call beans are really the seeds of plants. These seeds come packed in a "suitcase," called a pod. Sometimes we eat the pod and the bean seeds (string beans); other times we "unpack" the pod and just eat the seeds (lima beans). Beans are not the only plants that package their seeds in a pod. Peas do, too. Share the book *Bean* with your students, and watch their excitement grow as they undertake the investigation described below.

DIRECTIONS

1 Divide the class into groups. Give each group magnifying glasses and a paper plate containing samples of fresh and dried beans and peas.

2 Ask students to sort the beans and peas into those with a seam and those without.

3 Invite students to use a variety of methods to open the fresh pods and inspect the seeds inside. Discuss which method for opening the pods worked best for them and which were easiest to open.

4 Ask students to sort the seeds in other ways (e.g., by size, shape, color, or texture). Discuss their results.

5 As an extension, have students glue assorted dried beans and peas to construction paper to create a design. Display the designs on a bulletin board.

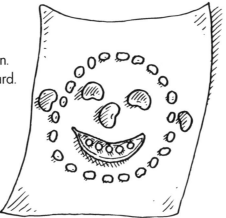

MATERIALS

◆ *Bean*
◆ magnifying glasses
◆ paper plates
◆ fresh and dried beans and peas
◆ glue (optional)
◆ construction paper (optional)

IN YOUR JOURNAL

1. What kind of beans did you examine? How were they similar and different? Which beans had a seam?
2. How did you open the beans?
3. How did you sort the beans?
4. List as many types of beans as you can. Circle the ones you have eaten.

BEAN SPROUTS

You do not always have to wait for a bean plant to complete its life cycle before you can use it for food. Many people eat bean sprouts. After bean seeds are well moistened, they begin to sprout. The bulk of the bean seed is stored food that the sprout uses until it is able to make its own food. Bean sprouts can be eaten raw or cooked. They are commonly used in salads, stir-fry dishes, and egg rolls. The most frequently eaten bean sprouts come from mung beans. After reading the book *Bean* to your students, invite them to take a closer look at some bean sprouts. Then, they can have some fun eating the results of their bean sprout experiments!

DIRECTIONS

1 Divide the class into small groups. Give each group magnifying glasses and a paper plate containing fresh bean sprouts.

2 Ask students to carefully observe the beans sprouts and describe what they see. Discuss the parts of a bean seed and sprout (the root, new leaves, and seed leaves, or cotyledons), and help students find them.

3 Ask students if they have ever eaten bean sprouts. Wash another sample of fresh bean sprouts, drain them on a paper towel, and place them on a paper plate. Pass the plate around, and encourage students to taste the bean sprouts and share their opinions about the taste and texture.

4 As an extension, have students work together to make a bean sprout salad. Invite them to taste the salad and discuss how they like it.

MATERIALS

- *Bean*
- magnifying glasses
- paper plates
- fresh bean sprouts
- paper towel
- salad ingredients: sprouts, celery, green onion, sesame seeds, sweet-and-sour dressing (optional)

IN YOUR JOURNAL

1. What does a bean sprout look like? Describe one and draw a picture.
2. What would happen if you planted and watered a bean sprout?
3. What are some other plants you can eat?

COMPARING SEEDS

It is not hard to grow your own beans. You can watch the life cycle unfold right in front of you. It is exciting to see the young bean plants push up through the soil, open small green leaves, and finally develop into mature plants that produce their own bean pods filled with seeds. There are many different types of beans to choose from. You can grow beans in a garden or in flowerpots. Share the book *Bean* with your students to prepare them for the changes they will see take place as their own bean plants grow and flourish.

DIRECTIONS

1 Give each student one clear plastic cup for each type of seed to be planted.

2 Have students fill the cups almost full with potting soil.

3 Invite students to add water until the soil is moist and sticks together. (It should not be soupy.) This should lower the level of the soil in the cup.

4 Have students use their fingers to poke a hole about 2" (5 cm) deep in each cup's soil.

5 Give each student one of each type of seed. Ask students to drop a seed into each hole and cover the seeds with soil.

6 Ask students to attach a piece of masking tape to each cup and write their name and the type of seed on each piece of tape.

MATERIALS

- ◆ *Bean*
- ◆ clear plastic cups
- ◆ potting soil
- ◆ water
- ◆ assorted seeds (e.g., sunflower, corn, pinto bean, pea)
- ◆ masking tape
- ◆ graph paper
- ◆ rulers

7 Place the cups near a window.

8 Have students add water every few days until the seeds sprout above the surface of the soil. After that, they can add water as necessary.

9 Invite students to graph the growth of their seeds. Have them record the number of days until the seeds sprouted and use a ruler to measure the height of their plants as they grow. Invite students to compare the different seeds' growth rates and the appearance of their leaves. It may take several weeks for the plants to mature and produce bean pods of their own.

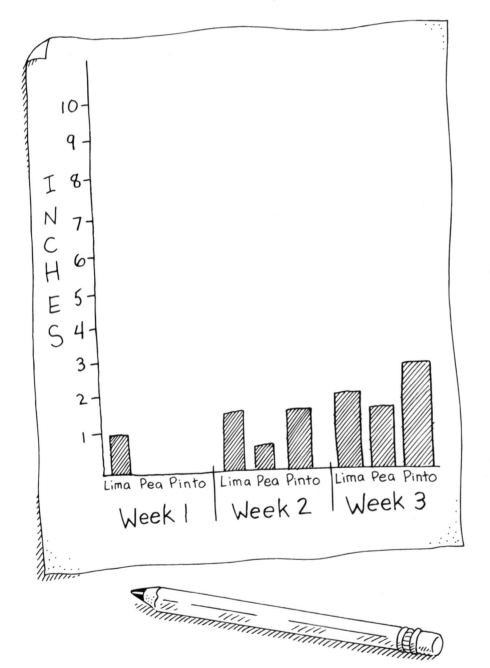

SUNFLOWER GAME

Sunflower plants need sunlight, water, and air in order to grow. Sunflower leaves spread out and reach for the sun. Their roots soak up the needed water. Carbon dioxide from the air enters through tiny holes (called stomata) on the underside of the leaves. As the plant matures, the giant flower head produces many seeds. The seeds, rich in protein and oil, are very tasty. Many animals and people like to eat these seeds. Read the book *Sunflower* with your students to acquaint them with the life cycle of a sunflower. Then, invite them to play the game described below to illustrate the life cycle steps.

DIRECTIONS

1 Make four copies of the Sunflower Game reproducible, and glue each copy to cardboard to make game cards. Color the background if you wish.

2 Use the plant parts on the reproducible as patterns to trace and cut from colored craft foam four brown root pieces, four green stems, four yellow flower heads, and 20 green leaves. Use a brown permanent marker to color the center of each flower shape.

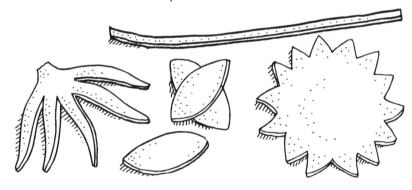

3 Use permanent markers to decorate a blank game die as follows: a yellow sun on two sides, a blue drop of water on one side, the word *air* on one side (or CO_2, if that is appropriate for your grade level), and a bug and/or caterpillar on the two remaining sides.

MATERIALS

♦ *Sunflower*
♦ Sunflower Game reproducible (page 17)
♦ glue
♦ 8 ½" x 11" (21.5 cm x 27.5 cm) cardboard
♦ crayons or markers
♦ scissors
♦ brown, green, and yellow craft foam
♦ permanent markers
♦ blank game die
♦ sunflower seeds
♦ box or tray

4 Discuss with students the steps of a sunflower's life cycle. Explain to them that they will have the opportunity to play a game in which the goal is to "grow" a complete sunflower plant.

5 Begin a demonstration of the game by giving two to four student volunteers each a game card and placing all the foam plant parts and 20 sunflower seeds in a box or tray.

6 Have players take turns rolling the die. If they roll something helpful (i.e., the sun, raindrop, or air), they place a foam plant part on their game board (starting at the bottom with the roots, then the stem, the leaves, the flower, and finally, the seeds). If they roll something harmful (i.e., the bug or caterpillar), they remove a part of their plant from their game board. If a player rolls the bug or caterpillar and has not yet built any part of the sunflower, he or she skips that turn.

7 Players take turns until one student has constructed a complete sunflower with roots, a stem, five leaves, a flower head, and five sunflower seeds placed onto the flower center.

8 After completing the demonstration, place the materials in a learning center, and invite students to play the game during scheduled center time or during their free time.

IN YOUR JOURNAL

1. What are the parts of a sunflower?
2. Draw a picture of a sunflower.
3. How can bugs and caterpillars harm sunflowers?
4. What do sunflowers need to grow?

Sunflower Game

MODEL SUNFLOWERS

Sunflowers are a very large type of composite flower (many small flowers compacted together). Your students can create their own sunflowers, first illustrating how the center is actually made of many small flowers called florets, and then showing how the florets in the center each contain a striped sunflower seed. Share the book *Sunflower* with your class, and invite them to create their own model sunflowers in the two-part activity described below.

DIRECTIONS

Part One: The Whole Flower

1 Share the book *Sunflower* with your class, paying particular attention to pages 1–9. Point out the vast number of florets found in the center, or head, of the sunflower.

2 Divide the class into groups, and give each group a paper plate, tissue-paper squares, pencils, construction paper, scissors, and glue.

3 Demonstrate to students how to wrap a tissue-paper square around the eraser end of a pencil and press or fold it upwards.

4 Demonstrate how to put a drop of glue on the plate and lower the tissue paper onto the glue. When the pencil is removed, the tissue-paper "floret" will be attached to the paper plate.

MATERIALS

- *Sunflower*
- paper plates
- 1¹/₂" (4 cm) brown tissue-paper squares
- pencils
- yellow and green construction paper
- scissors
- glue
- plastic knives
- round sugar cookies
- chocolate frosting
- candy corn
- yellow candy sprinkles

5 Ask students to attach sunflower florets all over their paper plate. (This step can be spread out over more than one day.)

6 Have students cut large petals from yellow construction paper and a stem and leaves from green construction paper.

7 Ask them to glue the stem and leaves to the bottom of the plate and the petals to the back of the plate to form several layers of yellow petals surrounding the center head of florets. Display the completed sunflowers on a bulletin board.

Part Two: Sweet Model Sunflowers

1 Give each student a plastic knife and a plate with a sugar cookie, chocolate frosting, candy corn, and yellow sprinkles.

2 Have students spread frosting on the top and sides of their cookie, cover the top with sprinkles, and attach candy corn to the sides. (The sprinkles represent the seeds and the candy corn represents the petals.)

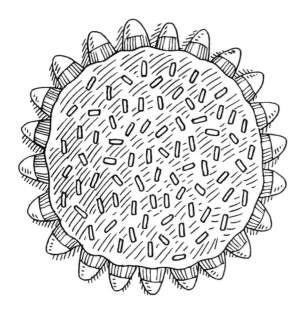

3 Invite students to snack on their sweet model sunflowers as you review with them the parts of a real sunflower.

IN YOUR JOURNAL

1. What is a floret?
2. Draw a diagram of a sunflower's flower head.
3. What is a composite flower?

GROW YOUR OWN SUNFLOWERS

Sunflower plants are not hard to grow. There are many varieties to choose from, some as small as 2' (61 cm) tall, and others that reach 8–10' (2.4–3.0 m). The most common varieties are yellow, but some sunflowers are orange or burgundy. Sunflowers can grow in almost any type of soil, and they are fairly drought resistant. They should be watered thoroughly once a week, especially during the first month. In North America and Europe, the best time to plant sunflowers is between March and September. In Asia, sunflowers grow best between February and November, and in Africa, sunflowers should be planted between November and May.

DIRECTIONS

1. Share the book *Sunflower* with your students to help them anticipate what will happen when they grow their own plants.

2. Fill a large planter box with soil, or have students loosen the soil of a small plot of land using gardening tools.

3. Water the soil until it is very moist.

4. Plant the seeds in a sunny location 1" (2.5 cm) deep and about 12" (30.5 cm) apart.

MATERIALS

♦ *Sunflower*
♦ large planter box or small plot of land
♦ soil
♦ gardening trowels or small shovels
♦ water
♦ sunflower seeds (from a nursery)
♦ large graph paper
♦ ruler
♦ salt water (optional)
♦ baking sheet (optional)
♦ oven (optional)

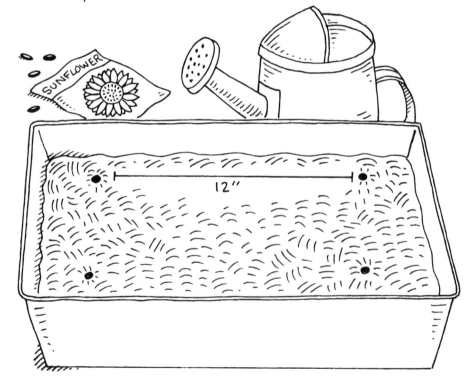

5 Have students water the seeds once a week, especially during the first month.

6 Invite students to graph the growth of the sunflower plants once a week. Depending on your weather conditions, the seeds should sprout in about a week.

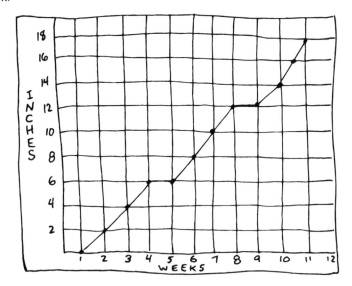

7 If you wish to harvest the sunflower seeds to culminate your sunflower study, wait for the seeds to form and dry out in the sunflower head. Hang the flower head upside down in a warm, dry place. Remove the dry seeds, soak them in a strong saltwater solution overnight, and drain them. Place the seeds on a baking sheet, and roast them for three hours at 200°F (93°C).

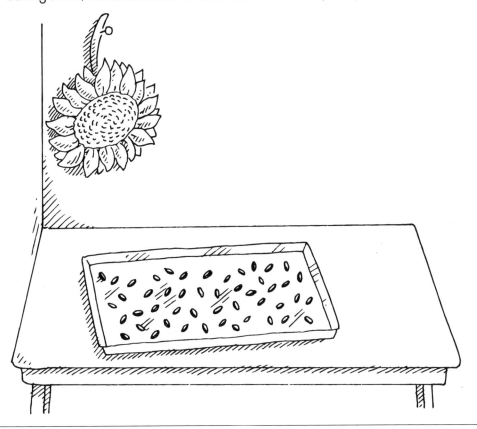

IN YOUR JOURNAL

1. How long do you think it will take the sunflowers to grow to full size? Why?
2. Why do sunflower seeds need to be watered more just after planting than after time passes?
3. What can you do to help sunflowers grow quickly?

WOOD FROG LIFE CYCLE WHEEL

There are many different types of frogs, but the stages of their life cycles are similar. All frogs begin as tiny eggs laid in water; hatch into tadpoles; undergo metamorphosis, developing legs and lungs to replace their tail and gills; and emerge as adult frogs. Some frogs, like most amphibians, spend time in or near water, but the wood frog only visits water for breeding. It spends most of its adult life on the floor of a woodland or forest area. Read the book *Wood Frog* to your students so they become familiar with this type of frog. Then, invite them to construct their own wood frog life cycle wheel to illustrate the changes that take place in the frog's life.

DIRECTIONS

1 Share the book *Wood Frog* while discussing with students the frog's metamorphosis and the coloration and dark eye mask typical of all wood frogs.

2 Give each student a card-stock copy of each Life Cycle Wheel reproducible. Have students draw a picture that matches the sentence(s) in each section of the second page and color both pages.

3 Have them cut out both parts of the life cycle wheel and use the tip of pointed scissors to poke a hole at the X on the top wheel and at the intersection of the lines on the bottom wheel.

4 Have students insert a brass fastener into the hole in the top circle and through the bottom circle and open the fastener flat on the back.

5 Have students write their name on the front of the wheel. Invite them to turn the wheel to observe the stages in the life cycle of a wood frog and read the stages to a partner.

MATERIALS

- *Wood Frog*
- card stock
- Life Cycle Wheel reproducibles (pages 23–24)
- crayons or markers
- scissors
- brass fasteners

IN YOUR JOURNAL

1. What are the stages of a wood frog's life cycle?
2. What is an amphibian?
3. Why is this frog called a "wood frog"?

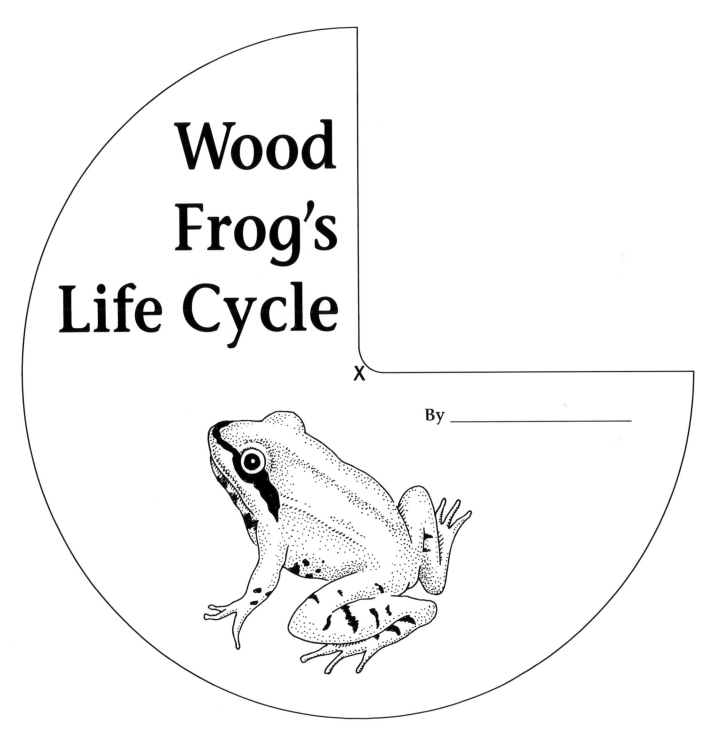

Wood Frog's Life Cycle

x

By _____

Life Cycle Wheel, Part 2

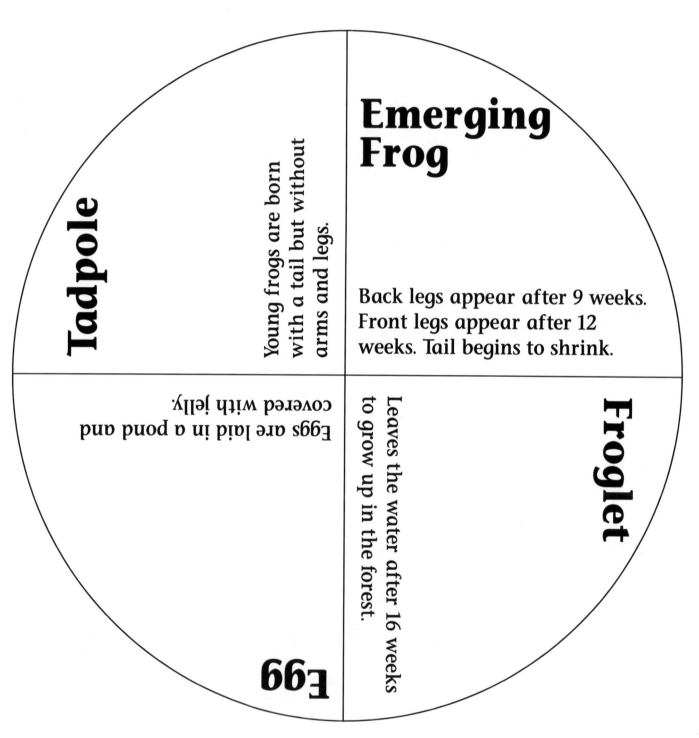

Tadpole

Young frogs are born with a tail but without arms and legs.

Emerging Frog

Back legs appear after 9 weeks. Front legs appear after 12 weeks. Tail begins to shrink.

Eggs are laid in a pond and covered with jelly.

Egg

Leaves the water after 16 weeks to grow up in the forest.

Froglet

SURVIVING A COLD WINTER

Wood frogs are the only North American frogs found north of the Arctic Circle. To survive the winter in very cold places, the wood frog stores extra sugar (glucose) in its liver as winter approaches. The glucose acts like antifreeze and protects the frog from cold. The frog digs a hole, buries itself, and hibernates through the winter. When the spring thaw comes, the frog slowly warms up and hops out of its burrow. Share the book *Wood Frog* with your students, and invite them to try the following experiment to see how the extra sugar in the frog's body keeps it from freezing to death during a harsh winter.

DIRECTIONS

1 Review the book *Wood Frog* with your students, and describe how the frog survives freezing cold winters.

2 Divide the class into groups, and give each group a mixing spoon, a bowl of water, a plastic cup filled with sugar, and two ice cube trays. Each group should have enough water to fill both trays.

3 Have group members take turns mixing water and sugar in their bowl, stirring and adding more sugar until no more will dissolve, creating a saturated solution.

4 Ask each group to fill one ice cube tray with the sugar-water solution, create a masking-tape label that reads *Wood Frog*, and attach the label to the tray.

5 Ask each group to fill their other tray with tap water, create a masking-tape label that reads *Tap Water*, and attach the label to the tray.

MATERIALS

- *Wood Frog*
- mixing spoons
- bowls
- water
- plastic cups
- sugar
- ice cube trays
- masking tape
- freezer
- paper towels
- pencils

6 Place each group's ice cube trays in a freezer. Check the trays two or three times after several-hour intervals to see if they are beginning to freeze. Invite students to record in their journal the time the water takes to start freezing or become hard on the top. Leave the trays in the freezer overnight. Check them again the next morning.

7 Divide the class into small groups. Remove the cubes from each tray, and give them to the groups. Have the groups place the cubes on paper towels labeled *Wood Frog* and *Tap Water*.

8 Invite students to observe the cubes, feel them, and poke them with a pencil. Ask students to describe the similarities and differences between the "wood frog" cubes and the tap-water cubes. Let the cubes start to melt, and challenge students to find which melt faster.

9 Discuss with students how the sugar-water ice cubes, which took longer to freeze, simulated how wood frogs keep from freezing to death in winter.

IN YOUR JOURNAL

1. How does a wood frog keep warm during winter?
2. Which tray of ice cubes froze faster? Why? Which melted faster? Why?
3. What does this experiment with ice tell you about wood frogs?

IF I WERE A FROG EGG

Frogs belong to a group of animals called amphibians. All amphibians undergo the process of metamorphosis, changing from a gill-breathing tadpole into a lung-breathing adult. There are a few amphibians that never lose their gills, even though they develop legs, but all amphibians start as tiny eggs in a mass of jelly floating in the water. Frog eggs have no shell to protect them and feel soft to the touch. The jelly that surrounds them offers some protection, but many frog eggs are eaten by predators even before they hatch. Share the book *Wood Frog* with your students, and let them investigate "frog eggs" in the activity below.

DIRECTIONS

(Note: Do steps 1–3 of this activity as a teacher-guided demonstration.)

1 Place 2 teaspoons (10 ml) of couscous in ¹/₂ cup (125 ml) of warm water, and let it soak for two minutes. Drain the water from the couscous with a strainer.

2 Crack a raw egg into a small glass dish or custard cup, and separate the white from the yolk. Discard the yolk. Use a mixing spoon to gently mix the couscous with the raw egg white. This mixture simulates frog eggs and the jelly in which they are hatched.

3 Have students observe the "egg and jelly" mixture. Invite students to take turns gently feeling the frog eggs and the jelly. Be sure students wash their hands after they finish.

4 Ask students to suggest words that describe the appearance, size, color, and texture of the frog eggs and the jelly. List the words on the board. Encourage each student to use the information in the book *Wood Frog* and the word list to create a haiku poem, a short story, or another form of creative writing about frog eggs. Create a bulletin board that features students' finished writing.

MATERIALS

- *Wood Frog*
- measuring spoon and cup
- couscous (semolina wheat grains)
- warm water
- strainer
- raw egg
- small glass dish or custard cup
- mixing spoon

IN YOUR JOURNAL

1. Describe the egg and jelly mixture.
2. Why do wood frogs lay their eggs in jelly?
3. What other materials could you use to represent frog eggs in jelly?

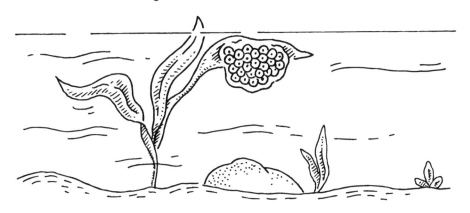

LADYBUGS UP CLOSE

There are about 4,000 species of ladybugs found all over the world, with over 350 species found in North America. The most common varieties of ladybugs are red with black spots and have a life cycle that lasts about one year. Ladybugs have a voracious appetite, and adults may eat 40–75 aphids a day. You can find ladybugs on rose bushes, dandelions, dill, and yarrow, or on wild grasses and bushes. They can also be purchased from plant nurseries. Read the book *Ladybug* to your students, and let them observe some ladybugs in your classroom to gain a better understanding of how these colorful creatures live.

DIRECTIONS

1 Place plant leaves into a terrarium or jar.

2 Add ladybugs, and replace the lid. Set the terrarium or jar on a table for viewing.

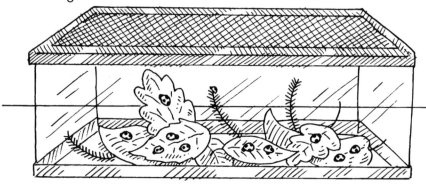

3 Have magnifying glasses and the book *Ladybug* available at the table.

4 Invite students to observe the ladybugs up close and compare them to the pictures in the book. Ask them questions such as *Are they the same variety? How many legs do you see? What do they do?*

5 Then, have each student paint a ladybug on the back side of a paper plate.

6 When the paint dries, invite students to write on the front of their plate a short story called *A Day in the Life of a Ladybug*, written from the ladybug's perspective. Hang the finished projects above students' desks.

(Note: If your leaves contain aphids, you can leave the terrarium set up for several days, adding new leaves with aphids as needed. If your leaves do not have any aphids, release the ladybugs after two days so they may find food.)

MATERIALS

- *Ladybug*
- ladybug plant leaves (e.g., rose, dandelion, dill)
- small terrarium or large jar with lid (with holes)
- ladybugs
- magnifying glasses
- paper plates
- black, yellow, orange, and red paint
- paintbrushes

IN YOUR JOURNAL

1. Compare the live ladybugs to the ladybugs in the book.
2. How can you help ladybugs survive in a terrarium?
3. List several things you observed while watching live ladybugs.

WHAT AN APPETITE!

The early stages of a ladybug's life cycle occur very quickly. Adult females lay clusters of 10–50 sticky, yellow eggs on the underside of a plant leaf where aphids live. The eggs hatch spiny larvae in about one week. The larvae eat a few hundred aphids over a period of two to three weeks before they undergo metamorphosis and emerge as an adult seven to ten days later. Read the book *Ladybug* to your students, paying particular attention to the early stages shown on pages 2–6. Then, let your class construct models of ladybug eggs and hungry larvae to help them learn more about the early stages of a ladybug's life cycle.

DIRECTIONS

1 Place a few handfuls of rice into a resealable plastic bag, and add several drops of yellow food coloring. Close the bag, and shake it until the rice is a pale yellow color. Store until ready to use. The grains of rice simulate ladybug eggs.

2 Divide the class into groups. (Each group will make one set of models.)

3 Give each group two pieces of green butcher paper, scissors, crayons or markers, and glue. Have students cut a large leaf (about as long as a desk) from each piece of paper. Ask them to draw veins on each leaf.

MATERIALS

- *Ladybug*
- rice
- resealable plastic bag
- yellow food coloring
- green butcher paper
- scissors
- crayons or markers
- glue
- card stock
- Larvae and Aphids reproducible (page 31)
- adhesive magnet strips
- transparent tape
- 2" (5 cm) black pipe-cleaner pieces

4 Give "ladybug eggs" to each group. Have students glue a cluster of eggs (10–50 pieces of rice) to the underside of one of their leaves. Display the leaves for a week while students complete the next steps of the activity—making ladybug larvae and aphids. (Remember, the eggs need one week to hatch.)

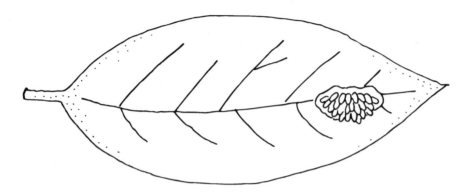

5 Give each group a card-stock copy of the Larvae and Aphids reproducible and several adhesive magnet strips. Have students color the aphids a green similar to the color of their leaves. Ask them to cut out the aphids and attach a small adhesive magnet strip to the underside of each one.

6 Invite students to color the ladybug larvae black with yellow spots. (Students can refer to the book *Ladybug*.) Have them cut out the larvae and attach a long adhesive magnet strip to the underside of each one. Then, have them tape six pipe-cleaner legs (three to a side) near the head of each larva. (Students can again refer to the book to see how the legs are attached near the larva's head.)

7 Encourage students to place their aphids all over their second green leaf. Then, have them move their larvae over the leaf and aphids. The aphids will disappear from the leaf as they are "eaten" by the hungry ladybug larvae! (The magnet strip on the back of the larvae will attract the small aphids and pick them up.) This process can be repeated so all students in the group get a chance to let a larva eat.

8 Discuss how many aphids a hungry larva can eat.

IN YOUR JOURNAL

1. What are the stages of a ladybug's life cycle?
2. In what ways is a ladybug larva different from an adult ladybug?
3. How can ladybugs be helpful to farmers and gardeners?

Larvae and Aphids

Larvae

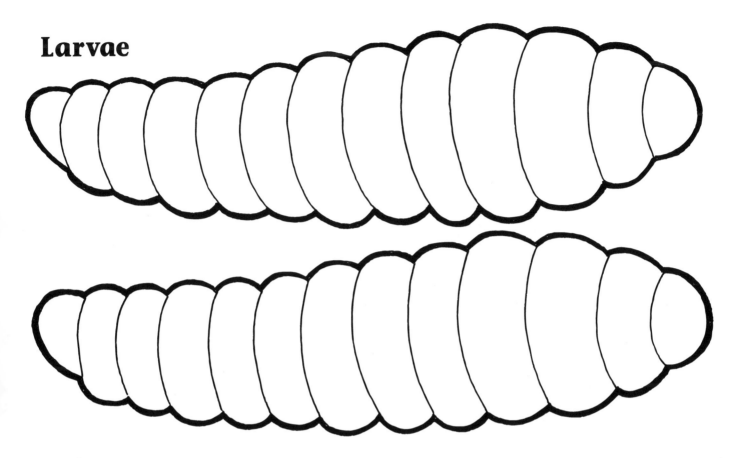

Aphids

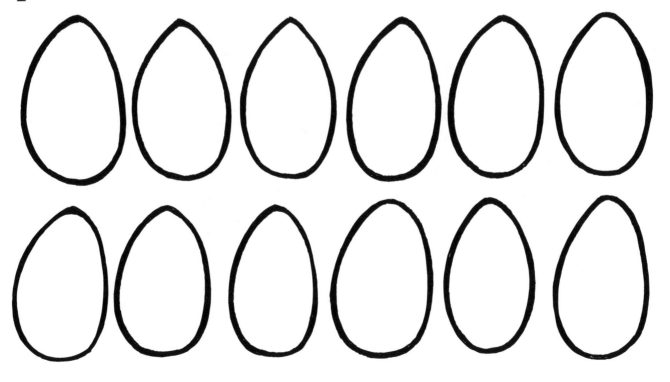

LADYBUG, LADYBUG, FLY AWAY HOME

Adult ladybugs are most often red, orange, or yellow with black spots. Different varieties have one, two, five, seven, eleven, or twelve spots. A few varieties have no spots, and one type is black with red spots. The colored part we see is actually a pair of hard wings that protect the ladybug. The bright colors warn birds that ladybugs do not taste good. Ladybugs can also "play dead" when in danger. Most predators will not eat an insect that is not moving. When a ladybug wants to fly, it opens its hard wings and uses a second pair of thin wings underneath. Read the book *Ladybug* to your students, and invite them to create their own adult ladybugs with spots, hard-shell wings, and delicate flying wings according to the procedures below.

DIRECTIONS

1 Give each student a copy of the Ladybug Patterns reproducible, and have students cut out the wing and body patterns.

2 Have students trace the body pattern on black construction paper and cut it out.

3 Ask students to trace the two wing patterns on tracing paper and cut them out.

4 Have students color the wing patterns red, orange, or yellow.

5 Invite them to draw black dots on the colored wings.

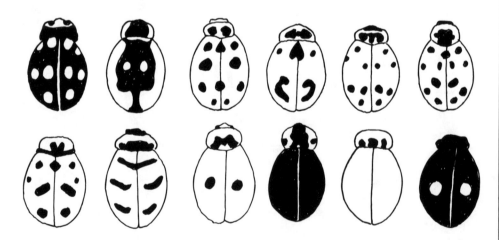

6 Have students place the tracing-paper wings on top of the black body and the colored wings on top of the tracing-paper wings. Show students how to use a scissors point to poke holes at the Xs on the colored wings and insert a brass fastener through all three papers. Have them open the fastener out flat on the bottom.

7 Have students write their name on the back of their ladybug.

8 Extend the activity by placing students' ladybugs at a learning center and inviting students to complete math activities such as sorting by the number of spots, adding the total number of spots, and writing story problems about ladybugs.

(Note: The ladybug is an insect and has six legs like all insects. This model emphasizes the two pairs of wings and how the ladybug flies. Legs have been left off this model for simplicity, but they could be cut from black construction paper and added, if desired.)

IN YOUR JOURNAL

1. How can you tell one type of ladybug from another?
2. Why does a ladybug have two sets of wings?
3. Describe differences between a ladybug's two sets of wings.

Ladybug Patterns

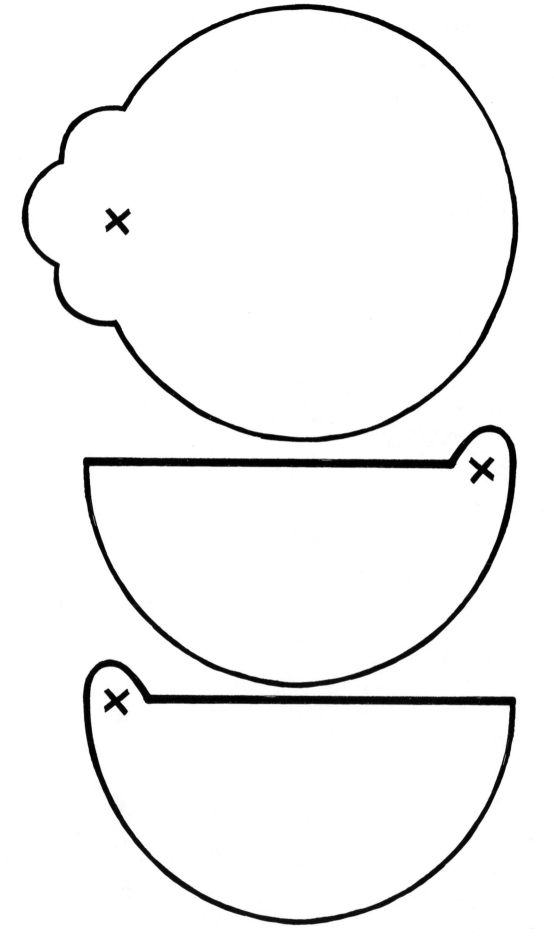

Body Pattern

Wing Patterns

Life Cycles © 1999 Creative Teaching Press

INVESTIGATING EGGS

A baby chick develops from the white of a fertilized egg. If an egg is not fertilized, the yolk (the chick's food supply) remains intact. Because grocery-store eggs are unfertilized, they still have a yolk and white part. Egg ranchers weigh and grade eggs before sending them to market. Small eggs weigh about 18 ounces (504 g) a dozen, medium eggs weigh 21 ounces (588 g) a dozen, large eggs weigh 24 ounces (672 g) a dozen, and extra-large eggs weigh 27 ounces (756 g) a dozen. "Grade AA" eggs have no cracks or defects in their shell and have a thick egg white. Eggshell color is determined by the color of the chicken. Share the book *Chicken* with your students, and then let them investigate a variety of eggs in the two-part activity described below.

DIRECTIONS

(Note: Have students wash their hands before and after the activity.)

Part One: Comparing Brown and White Eggs

1 Divide the class into groups. Give each group a white-shelled egg, a brown-shelled egg (preferably the same size), magnifying glasses, two small bowls or custard cups, and a balance scale.

2 Have students use the magnifying glasses to examine the eggshells. Discuss with students how the shells are alike and different.

3 Ask students to weigh the eggs on the balance scale and compare their findings.

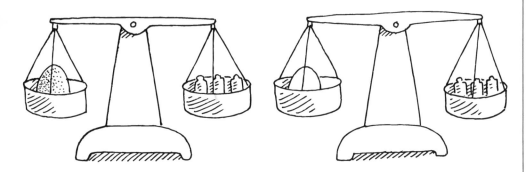

4 Have students gently crack open each egg over a small bowl or custard cup. Ask them to compare the white and the yolk to see if the eggs look the same inside.

MATERIALS

- *Chicken*
- small, medium, large, and extra-large white-shelled eggs
- brown-shelled eggs
- magnifying glasses
- small bowls or custard cups
- balance scales
- large graph paper
- glass measuring cups (optional)

Part Two: Size and Weight of Eggs

1 Give each group several eggs of different sizes.

2 Have groups weigh the eggs individually, or have each group weigh a different size of egg. Record their findings on large graph paper. (Calculate the average weight for each size, or graph the weights individually.) Ask students to compare the different weights. Challenge them to calculate how much a dozen of each size egg would weigh.

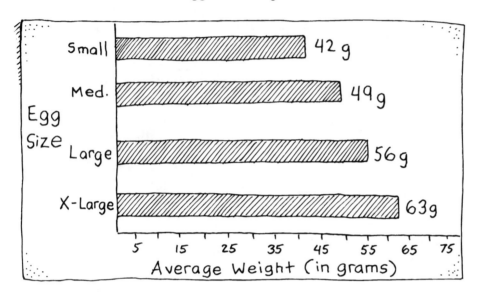

3 Gently crack a large egg into a small bowl or custard cup. Repeat for each size of egg you have.

4 Ask students to observe and describe the size of the yolk and the amount of egg white in each size of egg.

5 As an extension, give each group glass measuring cups. Challenge groups to find out how many eggs (small, medium, large, or extra large) it takes to measure $1/2$ cup (125 ml). Have each group measure a different size of egg.

IN YOUR JOURNAL

1. Describe what you learned about eggs.
2. What are some ways to sort chicken eggs?
3. Why won't baby chicks grow inside eggs you buy in the market?

I'VE GROWN MY COMB AND WATTLES

When a newborn chick hatches it is not very pretty. After several hours, the soft, downy yellow feathers dry and the chick looks cute and fluffy. At three to four weeks of age, a comb starts to grow on its head and mature feathers replace the down. By the time the chick is eight to ten weeks old, its comb is more pronounced and loose flaps of skin, called wattles, begin to form under its neck. When the chick is 14 weeks old, the comb and wattles are bright red and fully grown. Male chickens, called roosters, have larger combs and wattles than females. Read the book *Chicken* to your students. Then, invite them to create their own chicken life cycle models, from yellow chick to adult.

DIRECTIONS

1 Give each student a card-stock copy of the Chicken Patterns reproducible.

2 Ask students to cut out the pattern pieces. Have them trace the patterns on craft foam and cut out the pieces. (Have them use the color craft foam indicated on the reproducible.)

3 Have students use a black marker to add eyes or glue wiggly eyes on the chick and chicken.

4 Challenge students to assemble the pieces in the correct sequence and describe the life cycle of a chicken. They should start with the yellow baby chick, then place a small comb on top of the white chicken's head, and then exchange the small comb for a larger one and add the wattles under the chin to show an adult. Have them store their life cycle pieces in a resealable plastic bag.

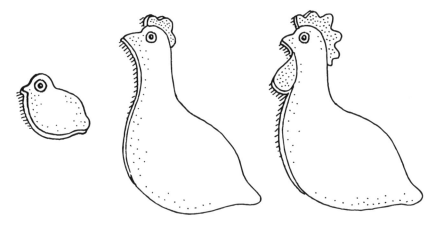

MATERIALS
- *Chicken*
- card stock
- Chicken Patterns reproducible (page 38)
- scissors
- yellow, white, and red craft foam
- black markers or wiggly eyes and glue
- large resealable plastic bags

IN YOUR JOURNAL
1. Describe the stages of a chicken's life cycle.
2. How can you tell an older chicken from a younger one?
3. What are a chicken's comb and wattles?

Chicken Patterns

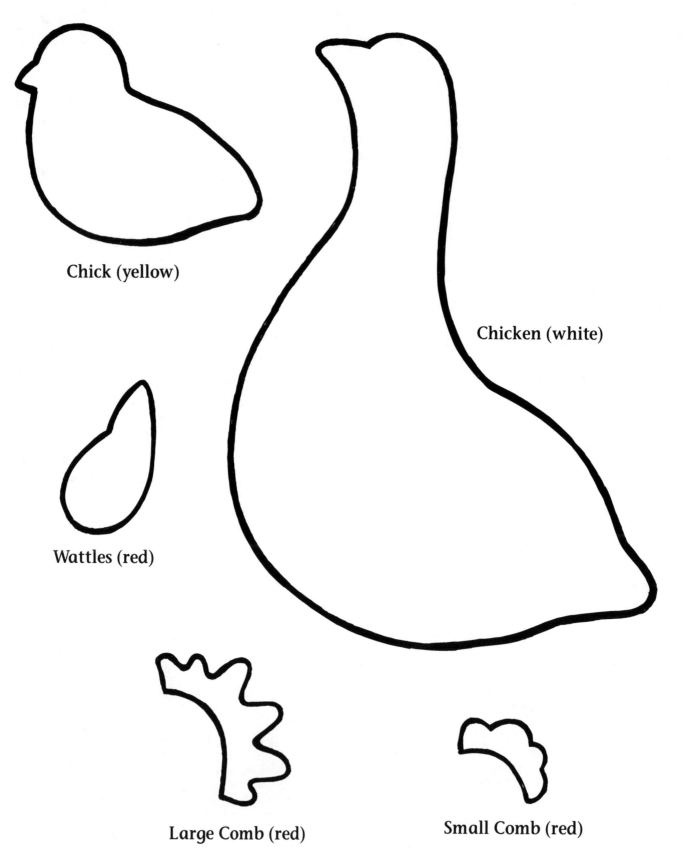

Chick (yellow)

Chicken (white)

Wattles (red)

Large Comb (red)

Small Comb (red)

SIMON CROWS

Chickens like to make their presence known. As little chicks, they make soft "cheep, cheep" sounds to let their mother know where they are. As they get older, their voices change, and they make "pawk, pawk" clucking sounds. Full-grown hens cluck and squawk even louder. The full-grown male rooster is the noisiest. Early in the morning, he loudly crows "Cock-a-doodle-do!" Read the book *Chicken* to your students, describing the sounds chickens make as they grow into hens or roosters. Challenge your students to give the appropriate sound when you hold up pictures of different stages in the life cycle of a chicken. Add excitement to the activity by playing Simon Says "chicken style," as described in the following activity.

DIRECTIONS

1 Photocopy the Noisy Chickens reproducible on card stock. Cut the pictures apart, and invite student volunteers to color them and return them to you.

2 Demonstrate for students the various sounds a chick makes as it grows up. Invite students to practice each sound.

STAGE	SOUND
baby chick	cheep, cheep
older chick	pawk, pawk
hen/rooster	cluck, cluck/cock-a-doodle-doo

3 Have students stand next to their desks. Stack the life cycle pictures in random order.

4 Play Simon Says chicken style by calling out the name of a stage as you show the picture, with and without saying *Simon says*. For more advanced students, say *Simon says* and hold up a picture without saying the name of the stage. (Students must recognize the stage in the life cycle and make the appropriate sound.) If you hold up a picture without saying *Simon says*, they should stay silent.

MATERIALS

- *Chicken*
- Noisy Chickens reproducible (page 40)
- card stock
- scissors
- crayons or markers

IN YOUR JOURNAL

1. Compare the sounds a chick makes to the sounds a chicken makes.
2. Why do you think baby animals make different noises than grown-up animals?
3. What other animals make different sounds when they are young than when they are old? What sounds do they make?

Noisy Chickens

Older Chick

Rooster

Baby Chick

Hen

WHAT IS A SPIDER?

Some people are afraid of spiders even though spiders seldom bite people. In fact, spiders can be helpful because they eat insects that can bother people or destroy plants. Spiders, part of a group of animals called arachnids, have two body parts, eight legs, eight eyes, and do not undergo metamorphosis like insects. A spider's legs are all attached to the head section of its body. A spider's body is protected by a hard outer covering, called an exoskeleton. This hard covering cannot grow or stretch, and the spider must shed this old skin and grow a bigger one several times during its life cycle. Share the book *Jumping Spider* with your students, and let them investigate more about spiders by having them construct their own model of a jumping spider.

DIRECTIONS

1 Use the pictures in the book *Jumping Spider* as a guide to discuss the structure of a spider's body. Make sure students understand that all eight legs are attached to the head section.

2 Invite students to choose whether they will make a male or a female spider. Give each student clay and eight pipe-cleaner pieces. (Have students use gray clay and black pipe cleaners for a male jumping spider and terracotta clay and brown pipe cleaners for a female.)

3 Have students make a spider head by shaping clay into a 1" (2.5 cm) ball. Have them make an abdomen by shaping another piece of clay into a slightly larger, flattened oval.

4 Have students bend their pipe-cleaner pieces into V-shaped spider legs and insert the legs into the head so the free ends rest on the table. Have them point some of the legs forward, some to the side, and some to the back.

MATERIALS
- *Jumping Spider*
- gray and terracotta colored clay
- 3" (7.5 cm) black and brown pipe-cleaner pieces
- toothpicks
- shoe boxes
- art supplies (e.g., raffia, craft sticks, tissue paper, paint)
- glue

5 Invite students to use a toothpick to poke eight eye holes in the jumping spider's head—four on the front and four on the top.

6 Invite students to use art supplies to create a shoe-box diorama of a spider's environment.

7 Have students glue the spider's head and body inside their shoe box. It is not necessary to glue the tips of the legs.

8 Have students write *Male Jumping Spider* or *Female Jumping Spider* and their name on the outside of the diorama. Display finished work on a shelf or large table.

IN YOUR JOURNAL

1. How is a spider different from an insect?
2. Describe and draw a diagram of the parts of a spider's body.
3. How do you think having eight legs helps a spider?

THE LIFE OF A SPIDER

There are thousands of types of spiders. The smallest is about the size of a letter "o" on this page, and the largest is about 7" (18 cm) long. Spiders catch their food in different ways. Many spiders spin a web to trap their prey. The white crab spider waits inside a flower for its food to come to it. The trapdoor spider digs a burrow in the ground and covers it with a trapdoor made out of spider silk. Wolf spiders chase their dinner. The jumping spider, featured in the book *Jumping Spider*, jumps on its prey. Share the book with your students, and encourage them to act out the life cycle of the jumping spider in the activity below.

DIRECTIONS

1 Review the book *Jumping Spider* with your students. Explain that they will be creating a skit based on the life cycle of a jumping spider.

2 As a class, brainstorm and list on the chalkboard the stages of the jumping spider's life cycle.

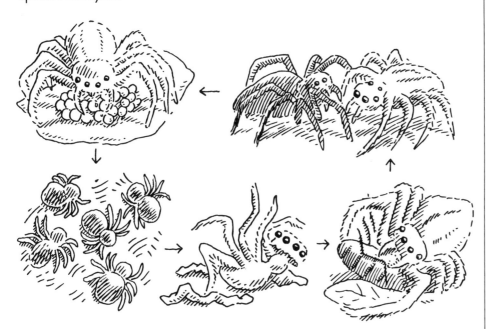

3 Divide the class into five groups.

4 Invite each group to select a stage in the spider's life cycle: an egg sac attached to a leaf, spiderlings leaving the egg sac, a jumping spider catching a fly, a spider shedding its skin, and a male courting a female.

5 Have the groups collect craft supplies to make stage props for their skit.

> ### MATERIALS
> - *Jumping Spider*
> - craft supplies (chosen by students)

6 Have groups plan how they will portray their stage in the cycle. The following are ideas for enacting some of the stages.

- Students in the "egg-sac" group can cut a big leaf shape from green butcher paper and tape it to the front of a small double row of chairs. One student acts as the mother spider and weaves a white yarn egg sac around the chairs. She then places the other members of her group (the eggs) inside the sac. They can each hold a picture of an egg in front of them.
- The "spiderling" group can create little spider pictures and some fly pictures for their costumes. Some can emerge from the sac and go searching for a fly to jump on.
- The "molting" group can have students remove sweaters and crawl away.
- The "courting" group can dress in black (males) with one member in brown (a female). The males "dance" in front of the female.

7 Invite the groups to take turns acting out the story of the jumping spider's life cycle in sequence.

IN YOUR JOURNAL

1. Describe the stages in a spider's life cycle.
2. How are newborn spiders different from newborn insects?
3. Describe how you acted out your life cycle stage.

SPIDER WEB MATH

Many spiders catch their prey in webs made of fine, silky threads. Spider webs come in various shapes and designs, including funnels, orbs, triangles, and domes. Some spiders just weave an irregular mass of silk threads. Web-spinning spiders usually construct their web and wait for an insect to crawl onto it. If a spider feels a very slight vibration on the web, it knows the vibration was probably made by wind or a drop of water. If the movement is very strong, the spider waits or hides, fearing an animal too big to catch is on the web. If the movement of the web silk feels just right, the spider investigates. Share the book *Jumping Spider* with your students, and invite them to play the math game described below in which the spider chases the fly.

DIRECTIONS

1. Fill copies of the Spider Web Math reproducible game board with addition, subtraction, multiplication, and/or division problems. The game can be used at a learning center or as a whole-class activity.

2. Laminate the game cards for durability.

3. Divide the class into pairs, and give each pair one game card and two game markers. One player, the "spider," places a marker in the middle of the web. The other player, the "fly," places a marker in any outer section of the web.

4. To play, students take turns selecting a math problem in a space adjacent to their current position. If they solve the problem, they move to that space. If not, they stay where they are. The object of the game is for the fly to cross the web before the spider catches it. Players can move in any direction on the web.

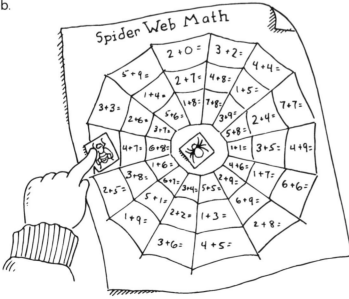

MATERIALS

- *Jumping Spider*
- Spider Web Math reproducible (page 46)
- laminator
- game markers (e.g., gaming chips, beans, plastic spider and fly)

IN YOUR JOURNAL

1. How does a spider know if there is prey on its web?
2. Describe different types of spider webs. Draw your favorite.
3. Tell about the game you played. What are some good strategies for the game?

Spider Web Math

BEAUTIFUL FALL LEAVES

Maple trees have broad, flat leaves with jagged edges along the lobes. In cold-weather regions, the leaves change colors in the fall and drop to the ground. Maple tree leaves contain pigments such as chlorophyll (green), xanthophyll (yellow), and carotene (orange). When there is more chlorophyll than other pigments, the leaves look green. As cold weather makes the chlorophyll disappear, red pigments called anthocyanins appear. This process gives us the varied and beautiful colors we associate with autumn. Share the book *Maple Tree* with your students, and invite them to create a maple tree filled with beautiful fall colors.

DIRECTIONS

1. Create a large tree trunk and branches from brown paper. Attach the pieces to a bulletin board.

2. Give each student a white construction-paper copy of the Maple Leaves reproducible and real maple leaves (if you have them). Invite students to observe the leaves' shape, edge design, and vein pattern.

3. Ask students to use a black crayon to draw veins on both sides of their leaves. Have them press hard on the crayon as they draw.

4. Invite students to paint both sides of their leaves shades of red, green, yellow, and orange. Have them cut out the leaves before the paint dries so the edges curl up like real leaves.

5. Help students attach their leaves to the bulletin-board tree branches.

6. As the class learns about the maple tree's life cycle, invite students to remove a few leaves at a time and attach them to the base of the tree, as if they had fallen off. Relate this process of change to a real maple tree.

MATERIALS

- *Maple Tree*
- scissors
- brown butcher paper or construction paper
- white construction paper
- Maple Leaves reproducible (page 48)
- maple leaves (optional)
- black crayons
- watercolors
- paintbrushes

IN YOUR JOURNAL

1. How does a maple tree change from season to season?
2. Why do maple tree leaves change color?
3. Draw a detailed picture of a maple leaf.

Maple Leaves

SPINNING SEEDS

In the spring, maple trees undergo a magnificent change. They develop small flowers that produce wing-tipped seeds called samaras. New green leaves appear on the tree branches. When summer comes, the wind blows the seeds off the tree branches. The seeds spin to the ground like little helicopters. This ability to "fly" helps the seeds disperse and land in a place where they can sprout. Then, the maple tree's life cycle can start all over again. Read the book *Maple Tree* with your students, and invite them to construct their own spinning-seed models.

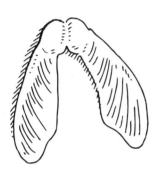  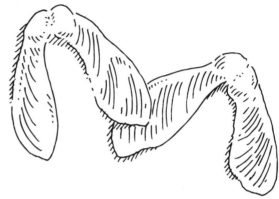

DIRECTIONS

1. Give each student a half of the Spinning Seed reproducible, and have students cut out the seed pattern.

2. Ask students to color their spinning seed (samara) brown on both sides. Be sure to point out where the two oval seeds would be on a real samara.

3. Have students fold the "wings" along the dotted lines—one forward, the other back.

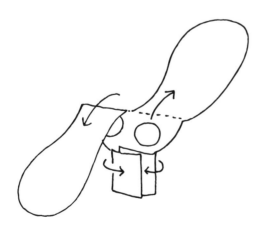

MATERIALS

- *Maple Tree*
- Spinning Seed reproducible (page 51)
- scissors
- brown crayons
- transparent tape
- small paper clips

4 Have them fold flaps A and B toward the center (one on top of the other) and tape the flaps down.

5 Ask students to attach a small paper clip onto the tab. (This provides the weight the seeds would give.) The paper clip should be in a vertical position on the tab.

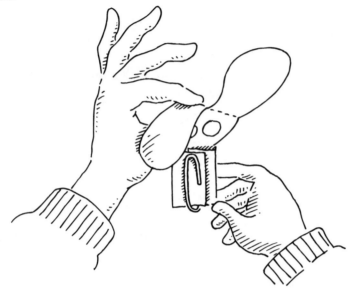

6 Invite students to hold their samara by the tab high above their head and release it. Ask them to observe the spinning motion as the samaras gently fall to the ground.

7 Explain that because of its wings, a samara can be blown far away from its maple tree.

IN YOUR JOURNAL

1. Why does a maple tree seed have wings?
2. What happened when you dropped your maple tree seed model?
3. Draw a picture of a maple tree seed.
4. List as many kinds of trees as you can.

Spinning Seed

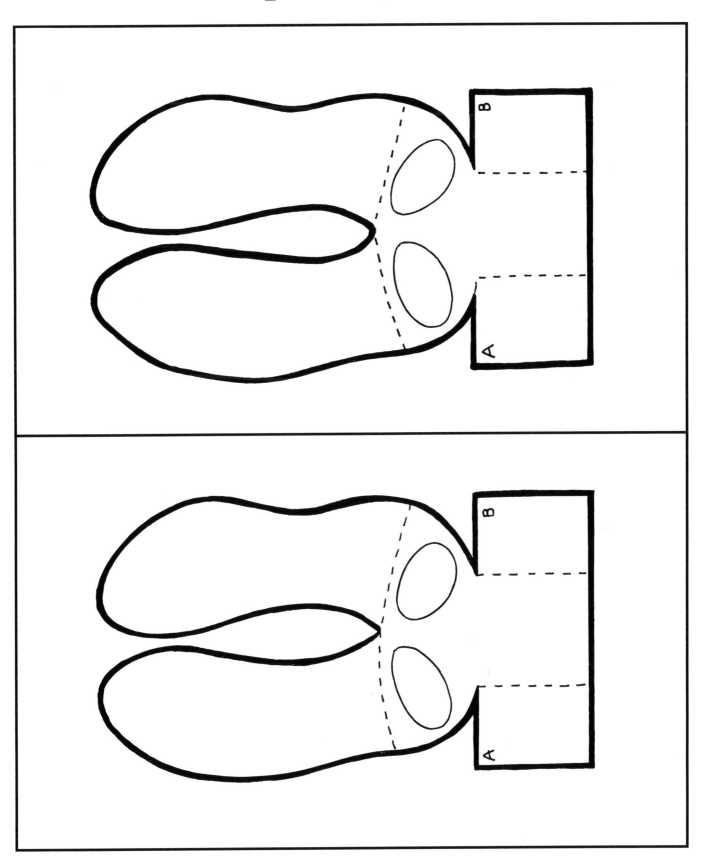

THE MAPLE TREE'S SWEET SAP

The maple family contains about 150 species of trees and shrubs. Some of the more common varieties are the sugar, black, red, silver, and big-leaf maples. Maple wood is used for making items such as furniture, bowling pins, flooring, and musical instruments. The sap of the sugar maple is used for making maple syrup and maple sugar. It takes 30–50 gallons (114–189 l) of sap to produce 1 gallon (3.8 l) of maple syrup. The sap begins flowing in early spring, when there are warm, sunny days and frosty nights. The sap harvest ends with the arrival of warm spring nights and the swelling of the maple flower buds. Share the book *Maple Tree* with your students, and invite them to investigate and compare maple syrup with other syrups in the activity described below.

DIRECTIONS

1 Have students observe pages 2 and 3 of the book *Maple Tree* to see how a maple tree looks around the time its sap is tapped. Discuss how real maple syrup is made from boiling down the sap. (Other syrups are usually made from cane sugar and water.)

2 Divide the class into small groups. Give each group masking tape and two pieces of waxed paper.

3 Have each group make two sets of masking-tape labels that read *Real Maple Syrup*, *Artificially Flavored Maple Syrup*, and *Molasses*. Have students place the labels across the top of both pieces of waxed paper, one set of labels per piece.

MATERIALS

- *Maple Tree*
- masking tape
- waxed paper
- toothpicks
- small paper cups
- real maple syrup
- artificially flavored maple syrup
- molasses
- maple cookie recipe and ingredients (optional)

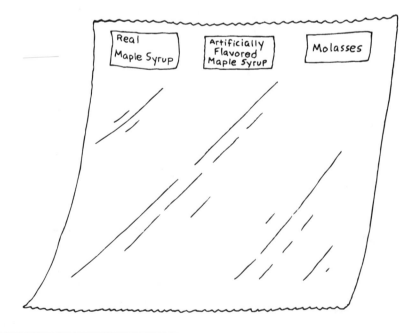

4 Give each group toothpicks and small cups of real maple syrup, artificially flavored maple syrup, and molasses. (Groups need only a small amount of each syrup.)

5 Ask students to use a toothpick to place a few drops of each type of syrup under its name on one piece of waxed paper.

6 Have students observe and describe the color and smell of each type of syrup. Then, invite them to use a clean finger to taste each type of syrup. Discuss their observations and preferences (e.g., best color, best smell, best taste).

7 Ask students to repeat Step 5 on the other piece of waxed paper.

8 Have students tilt the second waxed paper at about a 45° angle and observe which type of syrup runs down the paper fastest. Ask students which syrup is thickest and why they think so.

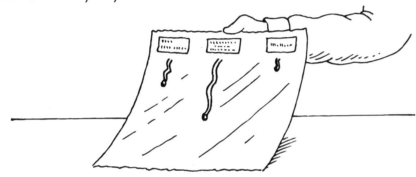

9 As an extension, invite students to make maple cookies (see recipe below). Make one batch with real maple syrup and one batch with artificially flavored maple syrup, and let students compare the taste.

┌───┐

Maple Cookie Recipe (makes two dozen)

$^1/_4$ cup (50 ml) butter $^1/_4$ cup (50 ml) raisins
$^1/_2$ cup (125 ml) maple syrup $^3/_4$ cup (175 ml) oatmeal
 1 egg $^1/_2$ tsp. (2 ml) nutmeg
$^3/_4$ cup (175 ml) flour $^1/_4$ cup (50 ml) milk
$^1/_2$ tsp. (2 ml) salt $^1/_3$ cup (67 ml) chopped nuts
 1 tsp. (5 ml) baking powder

Place butter, syrup, and egg in a bowl, and beat until light and creamy. Sift together flour, salt, and baking powder, and then fold into the liquid ingredients. Mix in raisins, oatmeal, and nutmeg. Add milk and nuts, and blend thoroughly. Drop spoonfuls onto a greased cookie sheet, and bake at 375°F (190°C) for about 15 minutes.

└───┘

IN YOUR JOURNAL

1. Describe the color, smell, and taste of each type of syrup.
2. Which type of syrup is thickest? How do you know?
3. How do you think people collect the sap from maple trees?

SIMULATING SNAKE EGGS

Snakes belong to a group of vertebrate animals called reptiles. They are covered with dry scales, breathe with lungs, are cold-blooded, and lay leathery eggs. These leathery eggs lack the calcium that hard-shelled bird eggs have. Some snakes build a simple nest of mud and leaves to protect their eggs, while many others just leave them uncovered in a secluded place. Read the book *Green Snake* to your students, and encourage them to try the experiment below to gain a better understanding of the differences between bird eggs and snake eggs.

DIRECTIONS

1 Divide the class into groups. Give each group two cups, two eggs, and some vinegar.

2 Have students gently place an egg into each plastic cup. Invite them to observe that the eggs look and feel the same.

3 Ask students to submerge one egg in vinegar and observe the bubbles forming on the shell as the vinegar dissolves the calcium carbonate in the shell. Discuss how this will make the eggs leathery, similar to a snake egg.

4 Set the eggs aside overnight.

5 Have students gently pour off the vinegar and remove their "snake egg." Invite them to rinse the egg with water if the vinegar smell is too strong. Ask them to place the egg on a paper towel to dry off.

6 Have students compare how the hard and soft eggs look and feel.

7 Have each student cut an egg shape from white construction paper and write on it a description of differences between the appearance and texture of their bird egg and snake egg. Display finished work on a science bulletin board.

MATERIALS
- *Green Snake*
- plastic cups
- raw chicken eggs
- vinegar
- paper towels
- scissors
- white construction paper

IN YOUR JOURNAL
1. How are snake eggs different from bird eggs?
2. If you were a baby animal, would you rather be inside a snake egg or a bird egg? Why?
3. Draw a diagram of a nest a snake could build to protect its eggs.

SNAKESKIN AND SCALES

Snakes, like other reptiles, are covered with dry scales. These scales form a diamond-patterned skin covering. Because the skin does not grow with the snake, it must shed its skin as it grows. The shed skin reveals the snake's scale pattern. This pattern varies from one type of snake to another. Contrary to popular belief, snakeskin is colorless. The colors that identify the snake are embedded in its body surface and are not lost upon skin shedding. Share the book *Green Snake* with your students, and have them create a flip book to demonstrate how a snake sheds its skin.

DIRECTIONS

1 If you have any real snakeskin, invite students to examine it and feel the scale pattern.

2 Give each student a copy of the Snakeskin reproducible, crayons or markers, and scissors.

3 Invite students to color the snake pictures, leaving the shed skin portions uncolored.

4 Have students cut the flip-book pages apart on the solid lines.

5 Ask students to stack the pages in order from page 1 to page 9 and staple the pages together along the top.

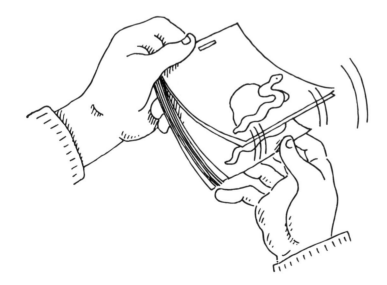

6 Invite students to flip the pages and watch as the snake sheds its skin.

MATERIALS

- *Green Snake*
- shed snakeskin pieces (optional, available from a pet shop)
- Snakeskin reproducible (page 56)
- crayons or markers
- scissors
- stapler

IN YOUR JOURNAL

1. Why does a snake shed its skin?
2. Describe how snakeskin feels.
3. How do snakes get their old skin off?

Snakeskin

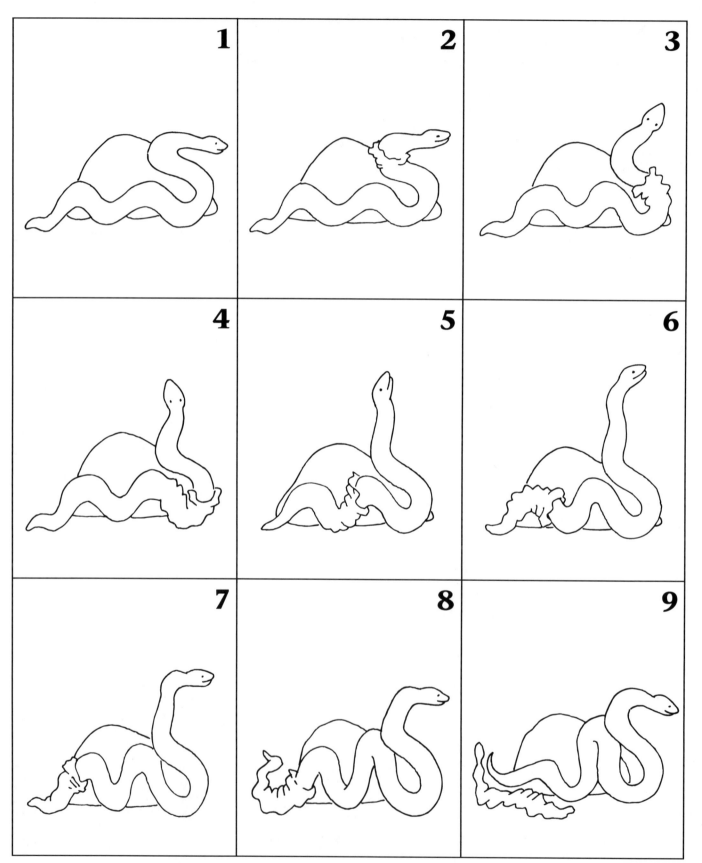

SKINNY AS A SNAKE

Snakes are very skinny animals. Because their bodies are so elongated, they are missing some body parts that other animals have. Snakes do not have outside ear openings or bladders, and many snakes only have one lung. They also lack eyelids, and, of course, snakes do not have any legs. However, snakes have an excellent sense of smell. They use their forked tongue and a special structure in the roof of their mouth, called the Jacobson's organ, to smell their surroundings and find their prey. Share the book *Green Snake* with your students, and invite them to create a model of a snake's skinny anatomy.

DIRECTIONS

(Note: You may wish to create a sample model students can refer to as they create their own.)

1 Give each student construction paper, crayons or markers, scissors, glue, and a copy of each Inside a Snake reproducible.

2 Ask students to color the internal organs of the snake.

3 Have students cut out parts 1, 2, and 3 of the snake's body pattern and arrange the pieces on construction paper so the ends marked A overlap and the ends marked B overlap.

4 Have them glue the three body parts to the construction paper.

5 Invite students to draw a forked tongue extending from the mouth. (Refer to page 13 of the book *Green Snake*.)

MATERIALS

- *Green Snake*
- large construction paper
- crayons or markers
- scissors
- glue
- Inside a Snake reproducibles (pages 59–60)

6 Have students cut out the organ pieces, and guide them through the following steps for completing their snake anatomy model. (Point out that the tip of the snake's tail does not contain any vital organs.)

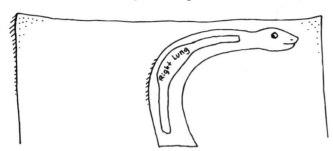

- Glue the right lung along the curve of the snake's neck.
- Glue the trachea/left lung and esophagus next to each other starting at the base of the head. The left lung and part of the esophagus should rest on top of the right lung.
- Glue the heart on top of the trachea and esophagus.
- Glue the stomach, the small intestine, and the large intestine to the end of the esophagus.
- Glue the liver next to the right lung.

7 Display finished projects on a bulletin board.

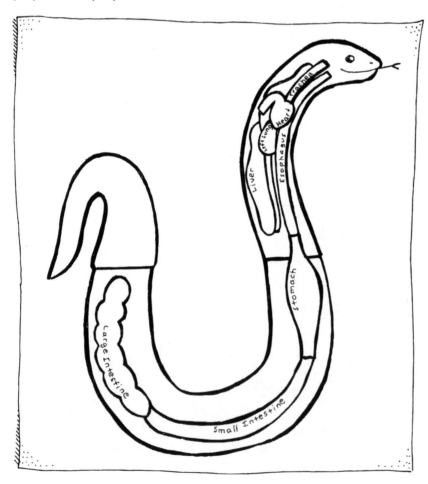

IN YOUR JOURNAL

1. How is the inside of a snake different from the inside of a person?
2. Many snakes can survive if their tail is cut off. Why do you think that is possible?
3. Why do many snakes have only one lung?

Inside a Snake, Part 1

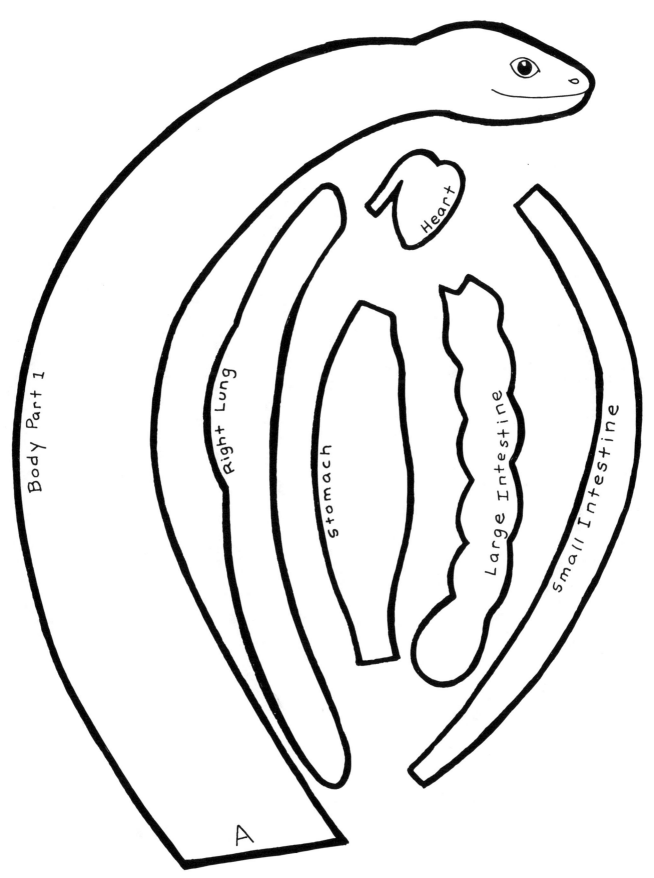

Inside a Snake, Part 2

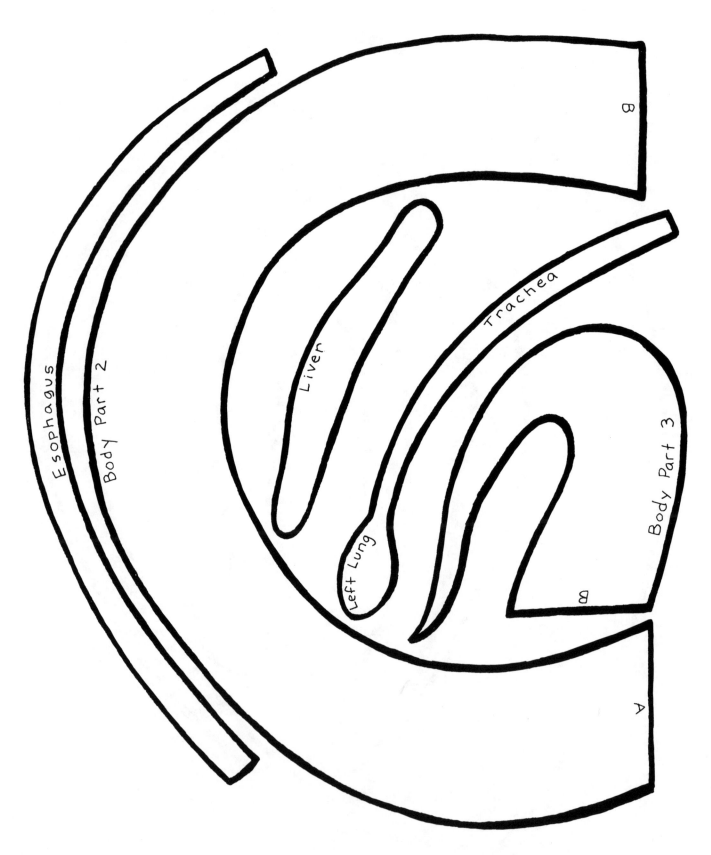

Life Cycles © 1999 Creative Teaching Press

TINY NESTS ARE BEST

Hummingbird nests are very tiny. The mother bird gathers bits of moss, spider silk, and leaves. She molds these materials together to form a cup-shaped nest about 1 ½" (4 cm) across. The opening to the nest is only about ³/₄" (2 cm) wide. When she is ready, she lays one or two tiny eggs inside the nest. The eggs are about the size of a raisin or miniature jelly bean. She sits on the nest for two to three weeks, keeping the eggs sheltered and warm until they are ready to hatch. Invite your students to read the book *Hummingbird* and create their own tiny hummingbird nest to better understand this miniature marvel.

DIRECTIONS

1 Prepare green salt dough ahead of time. (Make one batch for every 15–20 students.)

Salt Dough Recipe
(makes enough for 15–20 students)

1 cup (250 ml) flour
½ cup (125 ml) salt
1 cup (250 ml) water

1 tbsp. (15 ml) cooking oil
1 tsp. (5 ml) cream of tartar
green food coloring

Mix all ingredients in a pot. Cook and stir over medium heat until a loose ball forms. Remove from the pot, and knead until cool. Store in a resealable plastic bag.

MATERIALS

- *Hummingbird*
- green salt dough (see recipe)
- mixing bowls
- white thread
- paper plates
- green moss (from a plant nursery)
- leaves and/or grass clippings
- scissors
- miniature jelly beans
- glue
- twigs

2 Share the book *Hummingbird* with students, paying particular attention to pages 2–7.

3 Divide the class into four or five groups. Give each group a mixing bowl, white thread (representing spider silk), and a paper plate with moss and leaves and/or grass clippings on it.

4 Have students cut the thread into several pieces about the length of their little finger and cut or tear the moss and leaves/grass into small pieces. Ask them to put all the pieces into their bowl and mix them together.

5 Give each student a 1" (2.5 cm) ball of green salt dough. Have students put their dough into their group's mixing bowl and coat the dough with the leaf/moss/thread mixture. Encourage students to press and mold the dough to incorporate as much of the mixture as possible.

6 Have students shape the dough into a ball and insert their thumb to form a cup-shaped "nest."

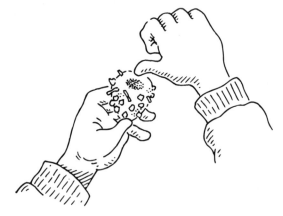

7 Give students one or two jelly-bean "eggs" for their nest. Invite students to glue the eggs inside the nest and glue their nest to a twig.

IN YOUR JOURNAL

1. What are some materials hummingbirds use to build nests?
2. Why do hummingbirds build cup-shaped nests?
3. When might it be helpful for a hummingbird to be small?

FLEDGLINGS LEARN TO FLY

Hummingbirds may be the smallest birds, but they are the fastest flappers. Birds like hawks and vultures can soar for long distances without flapping. Crows flap their wings about two times a second, and pigeons about three times a second. Adult hummingbirds flap their wings about 80 times each second. Their wings go so fast they make a humming sound. Hummingbirds are also the only birds that can fly backward and hover in one place. Imagine what a difficult time hummingbird fledglings have trying to leave the nest! Learning to fly like a hummingbird is not easy. It takes very strong chest and wing muscles. Read the book *Hummingbird* to your students, and invite them to try to fly like a hummingbird in the activity described below.

DIRECTIONS

1 Create on chart paper a large graph titled *How Fast Can You Flap?* Label three rows of the graph along the side *Crow, Pigeon,* and *Hummingbird Fledgling.* Label the bottom from zero to the number of students in the class.

MATERIALS

- *Hummingbird*
- chart paper
- watch with a second hand

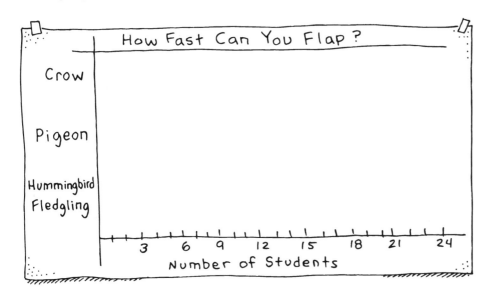

2 Discuss with students how fast a hummingbird flaps its wings compared to other birds.

3 Have students stand up with arms extended out from their sides. This is the position of a bird's wings when soaring.

4 Demonstrate to students how to flap their "wings" like a crow (two beats per second). Use a watch to time students for ten seconds as they count together while flapping their wings. (One up-and-down movement counts as one flap.) Graph the number of students who can flap like a crow (20 beats over ten seconds).

5 Demonstrate flapping like a pigeon (three beats per second), and graph the number of "pigeon flappers" in your class. Invite students to stop flapping if they get tired.

6 Ask students to pretend they are hummingbird fledglings and flap their wings as fast as they can while you time them for ten seconds. Remind them to stop if they get too tired. Graph the number of "fledgling flappers" in the group (students who could flap for the full ten seconds). Although they will not be able to flap 800 times during the ten seconds, they will get an idea of how fast a hummingbird's wings flap.

7 Discuss what happens to fledglings that cannot flap their wings fast enough or get tired. (They fall to the ground.) Refer to pages 8—9 of the book *Hummingbird.* Ask students to describe what it must be like to flap like an adult hummingbird. Ask which they think are stronger, a bird's wing muscles or a human's arm muscles and why they think so.

IN YOUR JOURNAL

1. Which are stronger, a bird's wing muscles or a human's arm muscles? Why do you think so?
2. Why do you think hummingbirds flap their wings faster than crows or pigeons?
3. Would you rather be able to fly like a crow or a hummingbird? Why?

PLANT A HUMMINGBIRD GARDEN

To support their busy lifestyle, hummingbirds need to eat about every ten minutes. Hummingbirds are attracted to the color and nectar of various shrubs and flowers, but they have a poor sense of smell, so they are not attracted to flowers that we think are fragrant. You can attract hummingbirds to your area by planting a hummingbird garden—a garden containing plants hummingbirds like to visit. Since hummingbirds are always on the lookout for something to eat, they will find your garden even if it is small. Share the book *Hummingbird* with your students, and invite them to plant a hummingbird garden. Then, invite them to study hummingbirds up close when the birds visit your garden.

DIRECTIONS

1. Divide the class into groups. Have each group work on a different part of the garden. Invite students to help clear the land, work the soil, fill planter boxes with soil, or plant flowers in the soil. (Be sure they wear gardening gloves for protection.) Consult your local plant nursery to see which hummingbird plants grow best in your area.

2. Have students take turns watering and caring for the plants.

3. Divide the class into groups, and assign each group a time to observe the garden. Have one group check the garden in the early morning as they arrive at school, another group check it at morning recess, another group check it at lunch recess, and a fourth group check the garden just before they leave school for the day. Have each group record how many hummingbirds they saw and which flowers the birds were near. Conduct these observations for a week or two.

4. Construct a class graph of these results to help students draw conclusions about which time of day their garden was most popular with hummingbirds and which flowers the birds preferred.

MATERIALS

- *Hummingbird*
- large planter boxes or small plot of land
- soil or planting mix (optional)
- plants (e.g., azalea, honeysuckle, yucca, columbine, lupine)
- water
- gardening gloves and tools
- large graph paper

IN YOUR JOURNAL

1. What did you see the hummingbirds in your garden do?
2. Why will hummingbirds visit certain types of flowers but not others?
3. What are other ways you might attract hummingbirds to your garden?

A FOAL STANDS UP

When a baby horse, or foal, is born, it can stand up when it is only an hour or so old. At first, the foal's legs are weak and wobbly. But after a few tries, the foal is much steadier on its feet. Human babies need many months of growth before they can stand up. Share the book *Horse* with your students, and suggest they try to imagine what it would be like to stand up so soon after birth. They can write their ideas in a pop-up book they create in the activity below and see the foal stand as it pops up in front of them.

DIRECTIONS

1 Cut construction paper into 5 ¹/₂" x 9" (13.5 cm x 23 cm) pieces. These are the covers for the pop-up books.

2 Share the book *Horse* with your students.

3 Give each student a copy of the Foal Pop-Up reproducible and a construction-paper book cover.

4 Ask students to write on the reproducible a short passage about a newborn foal trying to stand up. Invite students to write their passage from the foal's point of view.

MATERIALS

- *Horse*
- scissors
- construction paper
- Foal Pop-Up reproducible (page 68)
- crayons or markers
- glue

5 Invite students to color the foal on the reproducible. Have them cut out the pop-up page along the solid lines.

6 Have students fold the page forward along the dotted lines marked 1 and 2 to create a crease on each line. Ask them to open the page up again.

7 Ask students to fold the paper closed like a book along line 3 so the pop-up foal is folded inside the book.

8 Have students glue the construction-paper book cover to the back of the pop-up page. Invite them to write a title or draw a picture on the front and sign their name.

9 Invite students to open their pop-up book to read their story and watch the foal stand up.

10 Use these books during small-group story sharing and partner reading, or have students read their books to younger children in your school. During reading groups, invite students to pass their books around the group and read one another's stories.

IN YOUR JOURNAL

1. What would it be like to try to stand up only a few hours after you are born?
2. What are some things a foal and a human baby have in common?
3. How does the saying "If at first you don't succeed, try, try again" apply to a foal?

Foal Pop-Up

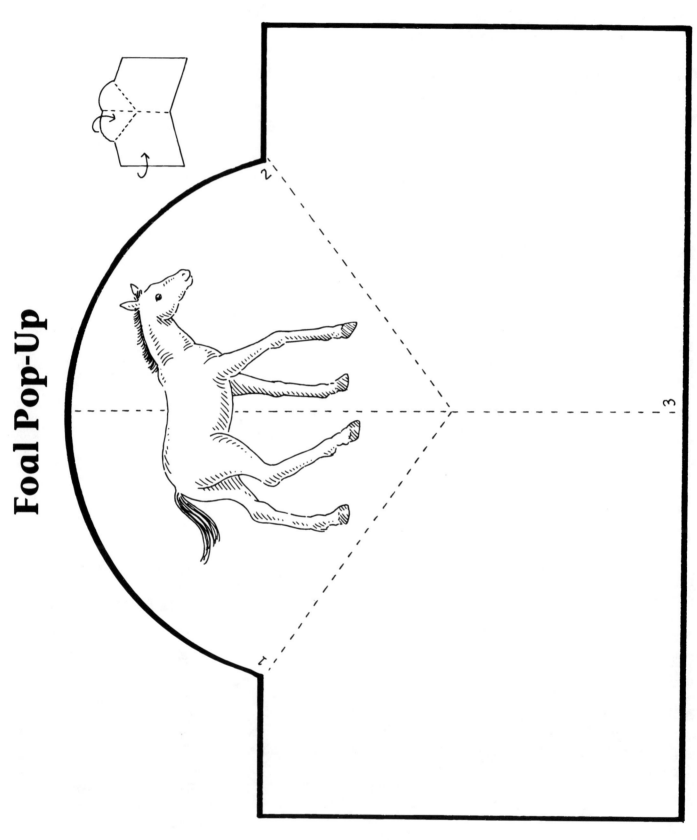

HORSE TAG

All young horses are called foals. Female foals are called fillies and male foals are called colts. All foals need lots of exercise. Right from the start, it is important for them to develop their leg muscles. Foals play with each other to get exercise. They run, kick up their heels, and chase each other. Horses make whinnying or neighing sounds. Invite your students to act like a young horse in the activity below. Read the book *Horse* to them so they can gather ideas about how young horses act. Then, take them outside for some "horseplay."

DIRECTIONS

1. Share the book *Horse* with your students, and demonstrate how to gallop like a horse.

2. Cut yarn into 2' (61 cm) pieces.

3. Make a "horsetail" by gathering a handful of yarn pieces together, wrapping one end of the bundle with another piece of yarn, and tying securely. Repeat so you have two horsetails.

4. Take students outside for a game of "horse tag." Choose one "filly" (girl) and one "colt" (boy) to be "it." Give them the yarn horsetails.

5. Tell students that they must gallop like a horse as they play. Invite them to make horse sounds while they gallop, if they wish. The students who are "it" try to tag another young horse with their yarn tail. When a student is tagged, he or she takes the horsetail and becomes "it." Allow enough time to play so that several students have a turn with a horsetail and can practice galloping and neighing. (Note: Be sure students tag each other gently with the horsetails.)

MATERIALS
◆ *Horse*
◆ yarn
◆ scissors

IN YOUR JOURNAL

1. What are male foals called? What are female foals called?
2. Why is it important for young horses to play together?
3. Which can you do faster, gallop or run? Why do you think that is?

A HORSE OF A DIFFERENT COLOR

Breeds of horses vary in size from small ponies to large draft horses. Some horses work on farms and ranches, others are bred for racing, and many are used for people's recreation. Some popular horse breeds are the Arabian, Morgan, Appaloosa, palomino, pinto, thoroughbred, quarter horse, and Clydesdale. All horses follow a similar life cycle. They develop inside the mother's body for about eleven months. A newborn foal grows rapidly. Most horses are fully grown in three to five years. Read the book *Horse* to your students, and invite them to learn about different breeds of horses. They can improve their observational skills by playing the horse match-up game described below.

DIRECTIONS

(Note: Cards for this memory match-up game can be made individually by students, or they can work together to make several sets for classroom use.)

1 Give each student or group crayons or markers and two card-stock copies of the Horse Cards reproducible. Read aloud and discuss with students the information on the Horse Guide reproducible.

2 Ask students to color the horse cards based on the information from the Horse Guide, making the two copies of the reproducible look the same.

3 Have students cut apart the cards.

4 To begin the game, have players turn the cards facedown and mix them up. Each player takes a turn turning over two cards, trying to find a match. If the cards match, the player keeps that pair. If not, he or she turns the cards facedown again. Play continues until all the cards are matched. The player with the most matched pairs wins.

MATERIALS

- *Horse*
- crayons or markers
- card stock
- Horse Cards reproducible (page 71)
- Horse Guide reproducible (page 72)
- scissors

IN YOUR JOURNAL

1. Which type of horse is your favorite? Why?
2. What are some ways people use horses?
3. Choose two types of horses and describe differences between them.

Horse Cards

Appaloosa	Arabian	Clydesdale
Morgan	Palomino	Pinto
Pony	Quarter Horse	Thoroughbred

Horse Guide

Appaloosa		a Spanish horse with spots that can cover the whole body, especially the face and rear flanks; spot patterns can appear to be a blanket, snowflakes, or splashes; some have striped hooves
Arabian		oldest and purest breed; about 15 hands high; slender legs and small feet; usually dappled gray, sometimes chestnut or bay (reddish brown)
Clydesdale		Scottish draft horse; about 17 hands high; very strong; large legs and feet; reddish brown with white blaze on the face and white lower forelegs; may be bay, brown, black, or chestnut
Morgan		often dark brown with black mane and tail; may have white diamond face marking
Palomino		tan to gold-colored quarter horse with very pale mane and tail; often have white feet and white blaze on face
Pinto		splotched pattern coloring, white/brown and white/bay common; mane may be darker than body splotches
Pony		smallest horses; less than 14 $\frac{1}{2}$ hands high; the smallest is the Shetland; come in a variety of colors, including white, tan, brown, and black
Quarter Horse		18th century North American breed; developed from thoroughbreds; agile with great endurance; mainly shades of brown; often have white feet and face markings
Thoroughbred		18th century English breed; about 16 hands high; small head and long neck; very fast runner; colors vary (most common are shades of brown)

Life Cycles © 1999 Creative Teaching Press

BUBBLE NESTS FOR BABIES

Unlike most fish, fighting fish can gulp air from the surface of the water. They have a special breathing organ, called a labyrinth, located behind their head that allows them to obtain oxygen from the surface. This feature helps them survive in their native habitat, the shallow water of rice paddies in Southeast Asia. When a male fighting fish is ready to mate, he swims to the surface and builds a nest of small bubbles. After mating, the male blows the eggs into the nest. He guards the nest for a few days until the babies, called fry, are ready to leave and live on their own. Share the book *Fighting Fish* with your students. Then, invite them to create bubble nests while investigating bubble-blowing techniques in the activity that follows.

DIRECTIONS

1 Prepare the bubble mix ahead of time, or let each group prepare their own.

> ### Bubble Mix Recipe
> (makes enough for one student group)
>
> 2 tbsp. (30 ml) liquid   6 tbsp. (90 ml) water
> dishwashing soap   4 tbsp. (60 ml) glycerin
>
> Gently mix liquid dishwashing soap, water, and glycerin in a mixing bowl. (Glycerin makes the bubbles much stronger and simulates the sticky saliva of the fighting fish.)

2 Divide the class into groups of two or three. Give each group a pie plate and each group member a straw.

3 Fill each pie plate with the bubble mixture. Challenge students to blow a mound of bubbles in the pie plate. This is their bubble nest.

4 Discuss with students bubble-blowing techniques they used to make their bubble nest. Ask students if blowing small bubbles was easier than blowing large ones. Invite them to discuss how they got their bubbles to stick together without floating away from the nest. Ask them which they think would be better protection for fighting fish eggs, big bubbles or small bubbles.

MATERIALS

♦ *Fighting Fish*
♦ bubble mix (see recipe)
♦ pie plates
♦ straws

IN YOUR JOURNAL

1. How are fighting fish different from other fish?
2. Why does a fighting fish build a bubble nest?
3. How did you build a bubble nest? What might have been a better way?

I'M JUST A SMALL FRY

Fighting fish babies, called fry, are tiny and translucent. Each fry has a yolk sac attached to its belly that provides it with food for the first few days after hatching. Then, the small fry must find food and avoid being eaten by predators. When they are about five weeks old, they develop an organ called a labyrinth that allows them to breathe air at the surface of the water, in addition to using their gills. They continue to grow until they are about one year old, developing larger fins and beautiful colors. Read the book *Fighting Fish* to your class, and invite your students to sing "I'm a Little Fish Fry" and act out the early parts of the fighting fish life cycle.

DIRECTIONS

1 Divide the class into three groups. Give each group tracing paper, craft sticks, tape, and scissors.

2 Have one group cut bubble shapes from tracing paper and tape the bubbles onto craft sticks. (These will make the nest for the fry.)

3 Give another group copies of the Fish Fry Patterns reproducible, and ask them to use the young fry pattern and tracing paper to draw and cut out two or three samples. Have them tape a craft-stick handle to each fry.

4 Invite the third group to use the older fry pattern to create two or three samples, cut them out, and tape a craft-stick handle onto the back of each one.

5 Gather the class at the front of the room. Place the bubble group in front, the young fry group in the second row, and the older fry group in the third row. (Not all students need a prop.)

6 Sing the song "I'm a Little Fish Fry" as a class. As students sing, have those with props hold them up to correspond with the lyrics. For example, the bubble group holds up their bubbles during the beginning line, the young fry group holds up their models during the next two lines, and the older fry group holds up their models during the last line. Invite students without props to perform the movements pictured on the song sheet.

MATERIALS

◆ *Fighting Fish*
◆ tracing paper
◆ craft sticks
◆ transparent tape
◆ scissors
◆ Fish Fry Patterns reproducible (page 75)
◆ I'm a Little Fish Fry song (page 76)

IN YOUR JOURNAL

1. What are the stages of a fighting fish's life cycle?
2. Describe the difference between a fighting fish fry and an adult fighting fish.
3. Why does a young fry have a yolk sac? What other animals have yolks?

Fish Fry Patterns

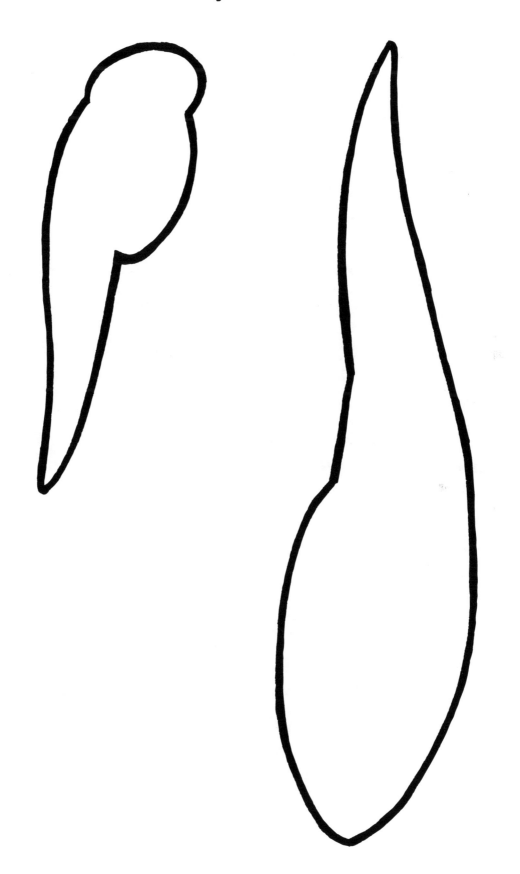

I'm a Little Fish Fry

(sing to the tune of "I'm a Little Teapot")

I'm a little fish fry, small and pale.

Here is my yolk sac. Here is my tail.

When I'm five weeks old, I'll change and grow.

Breathing air up top, I know.

DESIGN A BETTER BETTA

Fighting fish, also called bettas, develop beautiful fins and colors when they are adults. They may have one solid color, two colors, a "Cambodian" pattern, a "butterfly" pattern, or a "marbled" pattern. The betta featured in the book *Fighting Fish* has a solid-colored body. Invite your students to design their own betta based on the patterns that exist in nature. If it is convenient, purchase a betta at a pet shop, and invite students to observe the fighting fish firsthand. Bettas only need a small bowl of fresh water to swim in. Feed them freeze-dried blood worms, tropical-fish flake food, or brine shrimp. Change the water frequently.

DIRECTIONS

1. If you can obtain a live betta, set up the bowl for viewing in your classroom. Encourage students to help care for the fish.

2. Give each student a copy of the My Betta reproducible and coloring supplies.

3. Discuss with the class bettas' various colors and designs described below. Use colored chalk to draw each fish on the chalkboard as you discuss it.

4. Invite students to use your illustrations as a guide for designing and coloring their own betta.

5. Display the finished fish on a bulletin board with a fishbowl theme.

 **Solid:** entire body and fins are yellow, red, black, green, blue, turquoise, clear, or white

 **Two-colored:** entire body is one color and fins are a separate color (any of above colors)

 **Cambodian:** body is clear, white, or salmon colored, and fins are usually red, but may be any of the above colors

 **Butterfly:** body can be any of the betta colors, fins are partially colored in that color with clear or white tips; or body is colored and fins close to body are white, tips colored the same as body

 **Marbled:** face is white or salmon, body and fins are a splotched mix of two colors (green, red, blue, or steel blue)

MATERIALS

- *Fighting Fish*
- live betta, water bowl, fish food (optional)
- My Betta reproducible (page 78)
- colored chalk, oil pastels, or paint and paintbrushes

IN YOUR JOURNAL

1. Describe the betta that you created.
2. Which betta color pattern is your favorite? Why?
3. Why do you think different bettas have different color patterns?

My Betta

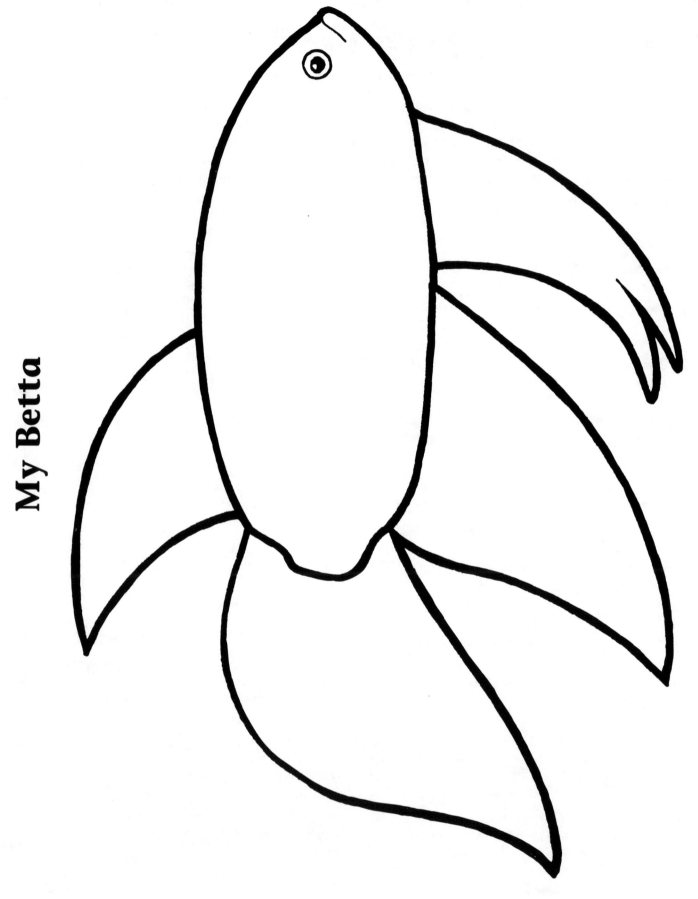

PUTTING IT TOGETHER

After students learn about the life cycles of the plants and animals covered in the *Life Cycles* series, conduct the following activities to help students recognize commonalities and interrelationships among all living things.

1 Invite students to compare life cycles in which the young resembles the mature adult to life cycles in which the young do not resemble the adult and, instead, undergo a metamorphosis (e.g., green snake or horse versus butterfly or wood frog). Have students divide the plants and animals from the *Life Cycles* books into these two categories and suggest other living things to add to each category.

2 Have each student research the life cycle of a plant or animal of his or her choice. Ask students to create a report and visual display, such as a mobile or poster, describing the stages of their plant or animal's life cycle. Encourage them to gather information from the Internet, encyclopedias on CD-ROM, and other sources where they can access photographs of their plant or animal.

3 Every animal species provides some sort of protection for its young. Some animal babies develop in an egg, others are hidden in a nest, and others stay with family members until they can care for themselves. Invite students to brainstorm how the nine animal species in the *Life Cycles* books provide protection for their young. Then, have them compare and contrast these methods. Invite them to select the animal they would most like to be, write why they chose that one, and describe the experiences they might have.

4 Plant seeds and animal eggs have many traits in common. As a class, brainstorm these traits and write them on index cards. Use hula hoops, jump ropes, or yarn to create a tabletop Venn diagram; place the cards and the diagram at a learning center; and invite students to place cards in the circles to compare seeds and eggs.

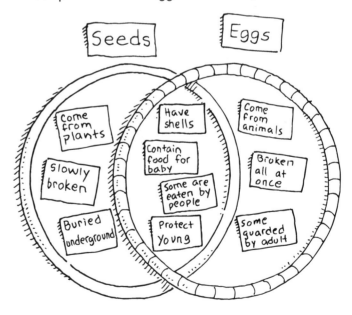

5 Place in a learning center a variety of animal and plant pictures cut from magazines. Challenge students to sort these pictures based on characteristics of the plants or animals. Depending on your students' ability level and your collection of pictures, you may wish to have them sort the pictures into the following categories:

- plants/animals
- birds/mammals/reptiles/fish/insects
- land animals/water animals
- plants/trees
- seeds/eggs/live babies

6 Characteristics common to all living things include the following:

- They grow and change.
- They take in nutrients.
- They give off waste.
- They reproduce.
- They are made of smaller structures.
- They respond to stimuli.

Review the animals and plants covered in the *Life Cycles* books, and challenge students to determine as many of these universal characteristics as possible on their own. Once the class has compiled the list, work as a group to create a chart that details how each characteristic pertains to each species. For example, discuss the sources from which hummingbirds, maple trees, and ladybugs get their nutrients.

7 Create a class big book about the life cycle of a human based on the format of the *Life Cycles* books. First, brainstorm the stages as a class. Then, invite students to suggest sentences to describe each stage. To complete the book, either divide the class into groups, and have each group use butcher paper to create pages for a stage in the cycle, or have each group create their own complete book. Invite students to use art supplies (e.g., construction paper, glue, paint, fabric, colored chalk) and pictures cut from magazines and printed advertisements to illustrate the pages. You may wish to provide students with the book's text written on sentence strips and have them glue the strips into the book.